Lecture Notes in Mathematics

Edited by A. Dold and B. Eckmann

1322

M. Métivier S. Watanabe (Eds.)

Stochastic Analysis

Proceedings of the Japanese-French Seminar
held in Paris, France, June 16–19, 1987

Springer-Verlag
Berlin Heidelberg New York London Paris Tokyo

Editors

Michel Métivier
Département de Mathématiques Appliquées
Ecole Polytechnique
91128 Palaiseau, France

Shinzo Watanabe
Department of Mathematics, Faculty of Science
Kyoto University, Kyoto, 606 Japan

Mathematics Subject Classification (1980): 60F, 60G, 60H, 60J

ISBN 3-540-19352-9 Springer-Verlag Berlin Heidelberg New York
ISBN 0-387-19352-9 Springer-Verlag New York Berlin Heidelberg

Printed in Germany

Printing and binding: Druckhaus Beltz, Hemsbach/Bergstr.
2146/3140-543210

PREFACE

As a project under the France-Japan Cooperative Science Program sponsored by C.N.R.S. (Centre National de la Recherche Scientifique) and J.S.P.S. (Japan Society for the Promotion of Science), a joint seminar on probability theory was held June 16-19, 1987, at Ecole Normale Supérieure, Paris. The main theme was stochastic analysis and applications to large scale systems. Nineteen lectures were presented on various topics like the Malliavin calculus, infinite dimensional stochastic differential equations and stochastic partial differential equations, limit theorems for particle systems, diffusions in random environment, hydrodynamical models, etc.

This volume of the Springer Lecture Notes is devoted to the original papers presented by the participants. A few lectures given at the seminar correspond to papers already published or being published elsewhere and are therefore absent from this volume. Because of the variety of the problems studied in those lectures, we did not find proper to try to group them -rather artificially- by topics, and adopted the alphabetic order of authors.

We would express our sincere thanks to contributors of this volume, all the participants of the seminar and also to Professor T. Hida who could not participate but, without whose kind advice and suggestions, this seminar could not have been realized.

It is also our pleasure to give our appreciation to Springer-Verlag for the prompt and efficient publication of the volume and Mrs Jeanne Bailleul for her help in the organization of the meeting and the preparation of the final volume.

February 5, 1988

Michel METIVIER - Shinzo WATANABE

<u>TABLE OF CONTENTS</u>

LIST OF PARTICIPANTS

D. BAKRY	Université Louis Pasteur, UER de Mathématiques 7, rue René Descartes. 67084 Strasbourg Cédex (France)
G. BEN AROUS	Centre de Mathématiques Appliquées. Ecole Normale Supérieure 45, rue d'Ulm. 75230 Paris Cédex 05 (France)
J.-M. BISMUT	UER 3e Cycle de Mathématiques. Université de Paris Sud Bâtiment 425. 91405 Orsay Cédex (France)
M. CHALEYAT-MAUREL	Laboratoire de Probabilités. Tour 56. Université de Paris VI 4, place Jussieu. 75252 Paris Cédex 05 (France)
N. EL KAROUI	Laboratoire de Probabilités. Tour 56. Université de Paris VI 4, place Jussieu. 75252 Paris Cédex 05 (France)
M. FUKUSHIMA	Department of Mathematics. College of General Education. Osaka University, Toyonaka, Osaka, 560 (Japan)
T. FUNAKI	Department of Mathematics. Faculty of Science. Nagoya University, Nagoya, 464 (Japan)
B. GAVEAU	UER 47. Laboratoire Analyse Complexe et Géométrie. Tour 45-46. Université Paris VI. 4, pl. Jussieu. 75252 Paris Cx 05 (France)
N. IKEDA	Department of Mathematics. Faculty of Science Osaka University, Toyonaka, Osaka, 560 (Japan)
K. ITO	RIMS, Kyoto University, Kyoto, 606 (Japan)
J. JACOD	Laboratoire de Probabilités. Tour 56. Université de Paris VI 4, place Jussieu. 75252 Paris Cédex 05 (France)
Y. KASAHARA	Institute of Mathematics. University of Tsukuba. Sakuramura. Ibaraki 305 (Japan)
C. KIPNIS	Centre de Mathématiques Appliquées. Ecole Polytechnique. 91128 Palaiseau Cédex (France)
S. KUSUOKA	RIMS. University of Kyoto, Kyoto, 606 (Japan)
R. LEANDRE	Département de Mathématiques. Faculté des Sciences de Besançon 25030 Besançon Cédex (France)
Y. LE JAN	Laboratoire de Probabilités. Tour 56. Université de Paris VI 4, place Jussieu. 75252 Paris Cédex 05 (France)
P. MALLIAVIN	10, rue Saint-Louis-en-l'Ile. 75004 Paris (France)
M. METIVIER	Centre de Mathématiques Appliquées. Ecole Polytechnique 91128 Palaiseau Cédex (France)
P.-A. MEYER	Institut de Recherche Mathématique Avancée. Rue du Général Zimmer. 67084 Strasbourg Cédex (France)

J. NEVEU — Laboratoire de Probabilités. Tour 56. Université de Paris VI
4, place Jussieu. 75252 Paris Cédex 05 (France)

A.-S. SZNITMAN — Courant Institute of Mathematical Sciences. New York University
251 Mercer Street. New York. N. Y. 10012 (U.S.A.)

H. TANAKA — Department of Mathematics. Faculty of Science and Technology
Keio University, Yokohama, 223 (Japan)

S. WATANABE — Department of Mathematics. Faculty of Science
Kyoto University. Kyoto, 606 (Japan)

TITLES OF LECTURES WHICH HAVE BEEN PUBLISHED SEPARATELY

C. KIPNIS and S. OLLA
Large deviations from the hydrodynamical limit for a system of independent Brownian particles.

J. NEVEU
Multiplicative martingales for spatial branching processes.

A.-S. SZNITMAN
Propagation of chaos for annihilating Brownian spheres.

NOYAU DE LA CHALEUR HYPOELLIPTIQUE ET GEOMETRIE SOUS-RIEMANNIENNE

Gérard BEN AROUS
Centre de Mathématiques Appliquées. Ecole Normale Supérieure
45, rue d'Ulm. 75230 Paris Cédex 05 (France)

I. INTRODUCTION

Nous allons décrire dans cet article les principaux résultats connus sur le comportement asymptotique du noyau de la chaleur associé à un opérateur elliptique dégénéré, illustrer les principaux phénomènes par des exemples et poser quelques problèmes encore ouverts.

Dans la suite, on considérera l'opérateur :

$$L = \frac{1}{2} \sum_{i=1}^{m} X_i^2 + X_o \tag{1.1}$$

où les X_i sont des champs de vecteurs C_b^∞ sur $\mathbb{R}^d$.

On fera toujours l'hypothèse de Hörmander forte :

$$\text{Lie}(X_1 \dots X_m)(x) = \mathbb{R}^d \qquad \forall x \in \mathbb{R}^d \tag{1.2}$$

qui assure que les opérateurs $\partial_t - L$ et L sont hypoelliptiques.

Soit alors $p_t(x,y)$ le noyau de la chaleur associé à L, c'est-à-dire la solution fondamentale de $\partial_t - L$, ou encore la densité de la loi de la diffusion $x_t(x)$ associée à L issue de x.

Cette diffusion est donnée par la solution de l'équation stochastique prise au sens de Stratonovitch :

$$\begin{cases} dx_t(x) = \sum_{i=1}^{m} X_i(x_t(x))dw_t^i + X_o(x_t(x))dt \\ x_o(x) = x, \end{cases} \tag{1.3}$$

où $(W_\cdot^i)_{1\le i\le m}$ désigne un mouvement brownien m-dimensionnel.

Nous voulons étudier ici le comportement asymptotique de $p_t(x,y)$ lorsque t tend vers zéro, ou encore le comportement asymptotique, lorsque ε tend vers zéro, de $p_1^\varepsilon(x,y)$ où $p_t^\varepsilon(x,y)$ est la densité de la loi de la solution $x_t^\varepsilon(x)$ de :

$$\begin{cases} dx_t^\varepsilon(x) = \varepsilon \sum_{i=1}^{m} X_i(x_t^\varepsilon(x))dw_t^i + \varepsilon^2 X_o(x_t^\varepsilon(x)dt \\ x_o^\varepsilon(x) = x. \end{cases} \tag{1.4}$$

En effet, par scaling évident, on a : $p_{\varepsilon^2 t}(x,y) = p_t^\varepsilon(x,y)$.

Nous noterons H^1 l'espace des fonctions à dérivées L^2 sur $[0,1]$ à valeurs dans $\mathbb{R}^m$, nulles en 0. Nous noterons $\mathring{h}$ la dérivée de $h \in H^1$ et $|h|_1$ la norme de h définie par :

$$|h|_1^2 = \sum_{i=1}^{m} \int_0^1 |\mathring{h}_s^i|^2 \, ds . \tag{1.5}$$

Considérons alors, pour $x \in \mathbb{R}^d$, l'application Φ^x qui, à un élément h de H^1 associe la solution $\varphi = \Phi^x(h)$ de l'équation :

$$\begin{cases} d\varphi_t = \sum_{i=1}^{m} X_i(\varphi_t)\, \mathring{h}_t^i \, dt \\ \varphi_0 = x . \end{cases} \tag{1.6}$$

On dira que $\varphi \in \Phi^x(H^1)$ est une courbe horizontale issue de x.

Pour x et y dans $\mathbb{R}^d$, considérons le sous-ensemble K_x^y de H^1 défini par :

$$K_x^y = \{h \in H^1 \,/\, \Phi_1^x(h) = y\} = (\Phi_1^x)^{-1}(\{y\}) . \tag{1.7}$$

Par l'hypothèse (1.2), cet ensemble est non vide ; on définit alors la distance sous-riemannienne (ou de Carnot-Carathéodory) par :

$$d^2(x,y) = \inf(|h|_1^2 \,,\, h \in K_x^y) . \tag{1.8}$$

L'infimum est atteint dans (1.8) et d définit une distance continue sur $\mathbb{R}^d \times \mathbb{R}^d$ (*cf.* [8], p. 33).

Si L est elliptique, cette distance coïncide avec la distance riemannienne associée à la métrique riemannienne définie par l'inverse du symbole principal de $2L$. Rappelons qu'alors L s'écrit :

$$L = \frac{1}{2}\Delta + X \tag{1.9}$$

où Δ est l'opérateur de Laplace-Beltrami pour cette métrique et X un champ de vecteur.

Le premier pas dans l'étude asymptotique de $p_t(x,y)$ a été fait par Léandre ([13]) et généralise le résultat de Varadhan [19] relatif au cas elliptique :

THÉORÈME 1.10.- *Avec les notations précédentes, on a :*

$$\lim_{t \to 0} t \log p_t(x,y) = -\frac{1}{2} d^2(x,y)$$

Remarque 1.11.- Ce résultat pourrait être obtenu de façon purement analytique en utilisant la technique introduite par Kannaï [11] dans le cas elliptique. C'est alors un résultat sur la vitesse de propagation des ondes pour l'opérateur $\partial_t^2 - L$.

II. DÉVELOPPEMENT ASYMPTOTIQUE DU NOYAU DE LA CHALEUR HORS DU CUT-LOCUS

1. Le cut-locus sous-riemannien

Nous allons ici donner le comportement asymptotique précis du noyau de la chaleur $p_t(x,y)$ pour les "bons points" (x,y) , i.e. les points hors du cut-locus. Commençons par définir cette notion de cut-locus qui généralise la notion correspondante dans le cas riemannien et qui a été introduite par Bismut [8].
Considérons le hamiltonien H associé à L , sur le cotangent de $\mathbb{R}^d$:

$$H(x,p) = \frac{1}{2} \sum_{i=1}^{m} \langle X_i(x),p\rangle^2 \quad \text{pour} \quad (x,p) \in T^*\mathbb{R}^d \tag{2.1}$$

et notons $\psi_t(x,p)$ le flot hamiltonien associé à H issu de (x,p) , et π la projection de $T^*\mathbb{R}^d$ sur $\mathbb{R}^d$.

DÉFINITION (2.2).- *Soient* x *et* y *deux points de* $\mathbb{R}^d$, *le couple* (x,y) *n'est pas dans le cut-locus si et seulement si :*

1) *Il existe un unique* $h_o \in K_x^y$ *tel que :* $|h_o|_1^2 = d^2(x,y)$.
2) *Il existe un unique* $p_o \in T^*\mathbb{R}^d$ *tel que :* $\Phi_t^x(h_o) = \pi\psi_t(x,p_o) \quad \forall t \in [0,1]$.
3) x *et* y *sont non conjugués, i.e. :* $\partial_p \pi\psi_1(x,.)_{|p=p_o}$ *est inversible.*

On dira que h_o *est la minimisante qui joint* x *à* h , *et que* $\Phi^x(h_o)$ *est la géodésique qui joint* x *à* y .

Remarque (2.3).- Cette définition coïncide avec la définition du cut-locus riemannien lorsque L est elliptique. En effet, dans ce cas, toute géodésique est projection d'une bicaractéristique et la condition 2) est toujours satisfaite. L'application qui, à $p \in T_x^*\mathbb{R}^d$ associe $\pi\psi_1(x,p)$ est alors simplement l'application exponentielle riemannienne usuelle $\exp_x$ (après identification de $T_x^*\mathbb{R}^d$ et de $T_x\mathbb{R}^d$).
Dans le cas où L est dégénéré, s'il est clair que la projection $\pi\psi_.(x,p)$ d'une bicaractéristique $\psi_.(x,p)$ est une courbe horizontale issue de x , il n'est pas évident par contre qu'une courbe horizontale issue de x (et en particulier la géodésique $\Phi_.^x(h_o)$) soit la projection d'une bicaractéristique. Le problème est de passer de la formulation lagrangienne du problème de minimum (1.8) à une formulation hamiltonienne ; c'est-à-dire d'être capable de définir un "état dual", ou encore d'écrire l'équation différentielle d'ordre 2 satisfaite par les géodésiques.

Les propriétés du cut-locus ainsi défini sont celles que l'on attend (*cf.* [8] et [3]) :

PROPOSITION (2.4).- 1) *Le cut-locus est fermé dans* $\mathbb{R}^d \times \mathbb{R}^d$.

2) $d^2(x,y)$ *est une fonction* C^∞ *de* (x,y) *sur le complémentaire du cut-locus.*

3) *L'unique minimisante* h_o *définie au* 1) *de la définition* (2.2) *varie de façon* C^∞ *en fonction de* x *et* y , *lorsque* x *et* y *varient dans le complémentaire du cut-locus.*

Remarque (2.5).- Dans le cas où L est elliptique, le complémentaire du cut-locus est un voisinage de la diagonale de $\mathbb{R}^d$ $\mathbb{R}^d$; ainsi deux points assez proches sont de "bons points". Ceci est faux si L n'est pas elliptique, précisément le cut-locus rencontre la diagonale de $\mathbb{R}^d \times \mathbb{R}^d$ exactement aux points (x,x) tels que L n'est pas elliptique en x.

Bismut a isolé ([8], th. 1.17) une condition suffisante (mais non nécessaire) pour qu'une courbe horizontale soit la projection d'une (unique) bicaractéristique :

PROPOSITION (2.6).- *Si l'application* Φ_1^x *de* H^1 *dans* $\mathbb{R}^d$ *est une submersion en* $h_o \in H^1$ (i.e. $d\Phi_1^x(h_o)(H^1) = \mathbb{R}^d$), *alors il existe un unique* $p_o \in T_x^*\mathbb{R}^d$ *tel que* : $\forall t \in [0,1]\ \pi\psi_t(x,p_o) = \Phi_t^x(h_o)$.

Il est facile de vérifier que cette condition de submersion est équivalente à la non-dégénérescence de la forme quadratique définie sur $T_x^*\mathbb{R}^d$ par :

$$<C^{h_o,x}p,p> = \sum_{i=1}^{m} \int_0^1 <\Phi_s(h_o)^{*-1}X_i(x),p>^2 ds \tag{2.7}$$

où $\Phi_s(h)$ désigne le flot de difféomorphismes défini par l'équation (1.6). $C^{h_o,x}$ est appelée matrice de Malliavin déterministe de h_o (en x).

Remarque (2.8).- On peut vérifier (*cf.* [3]) que dans la définition (2.2) du cut-locus, on peut remplacer la condition 2) par la condition 2 bis) suivante :

2 bis) Φ_1^x est une submersion en h_o .

La proposition (2.6) montre que 2 bis) implique 2); en fait, grâce à l'hypothèse 3) de la définition (2.2), 2) implique 2 bis).

En fait, on peut même affaiblir 2 bis) et la remplacer par :

2 ter) $d\Phi_1^x(h)$ est de rang localement constant au voisinage de h_o .

La condition 2 ter) est la condition naturelle pour assurer que l'ensemble $K_x^y = (\Phi_1^x)^{-1}\{y\}$ est une sous-variété hilbertienne de l'espace H^1 au voisinage de la minimisante h_o . Les points (x,y) tels que 2 ter) ne serait pas vérifié sont donc tels que l'ensemble des chemins K_x^y a une structure très singulière au voisinage de la minimisante h_o .

Un problème ouvert :

Nous mentionnons ici le fait que nous ne connaissons pas un seul exemple de points joints par une géodésique qui ne soit pas la projection d'une bicaractéristique. En effet, le contre-exemple fourni par [9] ne semble pas correct.

Signalons enfin que dans un article récent Strichartz [18] a donné une condition suffisante (la "strong bracket generating hypothesis") pour que toute géodésique soit projection d'une bicaractéristique. Cette condition contient la condition donnée par Bismut ([8], déf. 1.9) sous le nom d'hypothèse H2. La version publiée de [18] qui affirme trop rapidement que toute géodésique est projection d'une bicaractéristique sans aucune condition a en effet été récemment corrigée. Il serait très intéressant de savoir ce que deviennent les résultats très complets de Strichartz sans cette condition.

2. Le développement asymptotique du noyau de la chaleur

Pour des points (x,y) hors du cut-locus, le comportement du noyau de la chaleur en temps petit est analogue à celui du noyau de la chaleur elliptique. Cette conjecture, due à Bismut [8], a trouvé une réponse positive (dans [3] et [15]).

THÉORÈME (2.9) ([3]).- *Pour* (x,y) *hors du cut-locus, et* $n \geq 0$ *, on a le développement asymptotique à l'ordre* n :

$$p_t(x,y) = \frac{1}{t^{d/2}} e^{-\frac{d^2(x,y)}{2t}} \left(\sum_{j=0}^{n} c_j(x,y) t^j + t^{n+1} r_{n+1}(t,x,y) \right) \tag{2.10}$$

où :

a) *Les* c_j *sont des fonctions* C^∞ *sur le complémentaire du cut-locus.*

b) $c_0(x,y) > 0$.

c) *Pour tout compact* K *inclus dans le complémentaire du cut-locus, pour tous multi-indices* α, β *et tout entier* m *, il existe* $t_0 > 0$ *tel que*

$$\sup_{t \leq t_0} \sup_{x,y \in K} |\partial_x^\alpha \partial_y^\beta \partial_t^m r_{n+1}(t,x,y)| < \infty \ . \tag{2.11}$$

Ainsi le développement asymptotique (2.10) est uniforme sur les compacts qui ne rencontrent pas le cut-locus et la majoration (2.11) donne aussi évidemment un développement asymptotique de toutes les dérivées du noyau de la chaleur $\partial_x^\alpha \partial_y^\beta \partial_t^m p_t(x,y)$. Ce qui permet, grâce à l'équation satisfaite par $p_t(x,y)$, de vérifier que d est solution sur le complémentaire du cut-locus de l'équation de Hamilton - Jacobi $\sum_{i=1}^{m} (X_i d)^2 = 1$, et que les coefficients c_j sont solutions des équations de transport évidentes.

Nous allons donner ici une idée très rapide de la preuve du théorème (2.9) :
Soit F une fonction C^∞ bornée sur $\mathbb{R}^d$; par inversion de Fourier, on a :

$$p_1^\varepsilon(x,y)\, e^{-\frac{F(y)}{\varepsilon^2}} = \frac{1}{(2\pi)^d} \int d\xi\, E\left(e^{i\xi.(x_1^\varepsilon(x)-y)}\, e^{-\frac{F(x_1^\varepsilon(x))}{\varepsilon^2}} \right),$$

avec les notations de (1.4). Ou encore, après le changement de variable $\zeta = \varepsilon\xi$:

$$p_1^\varepsilon(x,y)\, e^{-\frac{F(y)}{\varepsilon^2}} = \frac{1}{(2\pi\varepsilon)^d} \int d\zeta\, E\left(e^{i\zeta.\left(\frac{x_1^\varepsilon - y}{\varepsilon}\right)}\, e^{-\frac{F(x_1^\varepsilon)}{\varepsilon^2}} \right).$$

Si on choisit F tel que l'infimum de $F + \frac{1}{2}d^2(x,.)$ soit atteint en y et que ce minimum soit non dégénéré (par exemple $F(z) = -\frac{1}{2}d^2(x,z) + |z-y|^2$) , les résultats de [6] sur la méthode de Laplace permettent d'évaluer asymptotiquement, à ζ fixé, l'espérance : $E\left(e^{i\zeta \frac{x^\varepsilon - y}{\varepsilon}}\, e^{-\frac{F(x_1^\varepsilon)}{\varepsilon^2}} \right)$.

En effet, dire que (x,y) n'est pas dans le cut-locus revient exactement à dire que les hypothèses variationnelles nécessaires pour utiliser les résultats de [6] sont

vérifiées dans cette situation très simple où la fonctionnelle $F(x_1^\varepsilon)$ ne dépend que du point final de la diffusion.
Enfin, le calcul de Malliavin permet d'intégrer en ζ ce développement asymptotique grâce à l'hypothèse de non-dégénérescence de la matrice de Malliavin déterministe (2.7). On obtient ainsi le développement cherché de $p_1^\varepsilon(x,y)$.

La nature de cette preuve est nouvelle, même dans le cas elliptique, son caractère essentiel étant d'utiliser directement les hypothèses variationnelles qui traduisent le fait que (x,y) n'est pas dans le cut-locus, à savoir que la géodésique réalise un minimum non dégénéré de l'énergie. En particulier, même dans le cas elliptique, elle a ceci de nouveau par rapport à la preuve de Molchanov [16] qu'elle n'utilise pas un résultat vrai pour des points proches, résultat que l'on propagerait aux points lointains grâce à la relation de semi-groupe. Ceci est important pour le cas dégénéré, car cette dernière approche n'est plus du tout possible dans ce cas si on veut un résultat fin sur $p_t(x,y)$.

3. Quelques exemples

a) *Le cas elliptique*

Si l'opérateur L est elliptique, le résultat du théorème (2.9) est bien connu (*cf*. [16], [1], [2], [8], [11] par exemple).
Rappelons simplement que, dans ce cas, la définition (2.2) coïncide avec celle de cut-locus riemannien usuel.

b) *Un exemple non-elliptique bien connu : le groupe d'Heisenberg*

Considérons, dans $\mathbb{R}^3$, les champs $X_1 = \partial_{x_1} - 2x_2\partial_{x_3}$ et $X_2 = \partial_{x_2} + 2x_1\partial_{x_3}$ et l'opérateur $L = \frac{1}{2}(X_1^2 + X_2^2)$.
Ces champs vérifient l'hypothèse H2 de Bismut, à savoir :

a) les champs X_i sont linéairement indépendants, et

b) pour tout $\lambda \in \mathbb{R}^2 \setminus \{0\}$, si on pose $Y = \lambda_1 X_1 + \lambda_2 X_2$, les champs : $X_1, X_2, [Y,X_1], [Y,X_2]$ engendrent $\mathbb{R}^3$ en tout point.

Ceci montre (*cf*. [8], th. 1.10) que, pour tout $h \in H^1 \setminus \{0\}$, la matrice de Malliavin déterministe $C^{h,x}$ est inversible et donc, par la proposition (2.6), que toute courbe horizontale non constante est projection d'une unique bicaractéristique. On peut calculer explicitement le flot hamiltonien (voir [9] et [2]), et on vérifie que le cut-locus est constitué des couples (x,y) tels que $x_i = y_i \quad i \in \{1,2\}$.
Ou encore, si on identifie $\mathbb{R}^3$ au groupe d'Heisenberg $\mathbb{H}$, et X_1, X_2 aux champs invariants à gauche qui engendrent l'algèbre de Lie de $\mathbb{H}$, le cut-locus est constitué des couples (x,y) avec $y-x$ dans le centre de $\mathbb{H}$.
Pour (x,y) hors du cut-locus, on obtient donc le développement asymptotique (2.10) qui est bien connu depuis que Gaveau [9] a montré comment il se déduit très simplement de la formule de l'aire de Paul Lévy.
Cet exemple, où L n'est elliptique nulle part, montre que le théorème (2.9) ne concerne pas que des cas "quasi-elliptiques". Cependant, on pourrait croire que

l'hypothèse H2 joue un rôle crucial, et en particulier que le théorème (2.9) ne s'applique que dans des cas où les crochets de longueur ≤ 2 des champs X_i suffisent pour engendrer l'espace tangent.
L'exemple suivant montre qu'il n'en est rien.

c) *Un exemple arbitrairement dégénéré*

L'exemple qui suit provient d'une discussion avec R. Léandre et sera développé dans un travail futur.
Considérons sur $\mathbb{R}^d$ (avec $d \geq 2$) les deux champs de vecteurs :

$$X_1 = \partial_{x_1} \quad \text{et} \quad X_2 = \sum_{i=2}^{m} x_1^{k_i} \partial_{x_i}$$

où $k_2 < k_3 < \ldots < k_d$ sont des entiers positifs.
L'opérateur $L = \frac{1}{2}(X_1^2 + X_2^2)$ satisfait à la condition de Hörmander forte (1.2), mais bien sûr L n'est nulle part elliptique si $d > 2$.
De plus, il est clair que la longueur des crochets des champs X_i nécessaires pour engendrer $\mathbb{R}^d$ tend vers l'infini avec la dimension d.
Soit, pour des réels positifs C,D,δ l'ensemble :

$$A(C,D,\delta) = \{(x,y) \in \mathbb{R}^d \times \mathbb{R}^d \ , \ C < |y_1 - x_1| < D \ , \ |y_i - x_i| < \delta \ \ \forall i \geq 2\} \ .$$

On a alors :

PROPOSITION (2.12).- *Pour tous réels positifs* $0 < C < D$, *il existe un* $\delta > 0$ *tel que* $A(C,D,\delta)$ *ne rencontre pas le cut-locus et donc tel que le développement asymptotique du noyau de la chaleur associé à* L *soit donné par le théorème* (2.9) *pour* $(x,y) \in A(C,D,\delta)$.

Preuve.- Le cut-locus étant fermé, il suffit de vérifier que, pour $(x,y) \in \mathbb{R}^d \times \mathbb{R}^d$ tel que : $x_1 \neq y_1$ et $x_i = y_i$ $i \geq 2$, alors (x,y) ne soit pas dans le cut-locus.
Soit donc un tel couple (x,y) et $h \in K_x^y$, on a alors $\Phi_t^x(h) = (x_i(t))_{1\leq i\leq d}$ avec :

$$\begin{cases} x_1(t) = x_1 + h_t^1 \\ x_i(t) = x_i + \displaystyle\int_0^t (h_s^1)^{k_i} dh_s^2 \ , \ \text{pour } i \geq 2 \ . \end{cases} \tag{2.13}$$

d'où : $|h|_1^2 \geq |h^1|_1^2 \geq |y_1 - x_1|^2$, ainsi $d^2(x,y) \geq |y_1 - x_1|^2$.
Mais en choisissant $h_s^1 = (y_1 - x_1)s$ et $h_s^2 \equiv 0$, on vérifie que $h \in K_x^y$ et que h est l'unique élément de K_x^y tel que : $|h|_1^2 = d^2(x,y)$, où bien sûr la distance hypoelliptique $d^2(x,y)$ est ici égale à la distance euclidienne :

$$d^2(x,y) = \|y - x\|^2 = |y_1 - x_1|^2 \tag{2.14}$$

Ainsi, le 1) de la définition (2.2) est vérifié.
D'autre part, le hamiltonien est ici égal à :

$$H(x,p) = \frac{1}{2}\left(p_1^2 + \left(\sum_{i=2}^{m} x_1^{k_i} p_i\right)^2\right) . \tag{2.15}$$

Le flot hamiltonien associé $\psi_t(x,p) = (x_t,p_t)$ est donné par :

$$\left\{\begin{array}{ll} \dfrac{dx_1}{dt} = p_1 & \dfrac{dp_1}{dt} = -\sum\limits_{i=2}^{d} k_i x_1^{k_i-1} p_i \\ \dfrac{dx_i}{dt} = \left(\sum\limits_{j=2}^{d} x_1^{k_j} p_j\right) x_1^{k_i} & \dfrac{dp_i}{dt} = 0 \qquad d \geq i \geq 2 . \end{array}\right. \tag{2.16}$$

On vérifie ainsi qu'il existe un unique $p_o \in T_x^*\mathbb{R}^d$ (à savoir $p_o = (y_1 - x_1, 0, \dots 0)$) tel que : $\forall t \in [0,1]$ $\Phi_t^x(t) = \pi\psi_t(x,p_o)$.

Le 2) de la définition (2.2) est donc vérifié.

Il reste à montrer que les points x et y sont non conjugués :

Le jacobien $\partial_p \pi\psi_1(x,.)_{|p=p_o}$ se calcule aisément par (2.16) :

$$\partial_p \pi\psi_1(x,.)_{|p=p_o} = \begin{pmatrix} 1 & * \cdots * \\ 0 & \\ \vdots & A \\ 0 & \end{pmatrix} \tag{2.17}$$

où la matrice $(d-1) \times (d-1)$ A est donnée par : $A = (a_{ij})_{1 \leq i,j \leq d-1}$ et

$$a_{ij} = \int_0^1 x_1(s)^{k_i+k_j} ds = \int_0^1 (x_1 + s(y_1 - x_1))^{k_i+k_j} ds . \tag{2.18}$$

Or A est inversible, comme matrice de Gram (pour le produit scalaire de L^2) des fonctions $s \longrightarrow x_1(s)^{k_i}$ $(2 \leq i \leq d)$. En effet, ces fonctions sont linéairement indépendantes du fait que les k_i sont distincts et que $x_1(s)$ n'est pas constante puisque $x_1 \neq y_1$.

Ainsi $\partial_p \pi\psi_1(x,.)_{|p=p_o}$ est inversible, ce qui montre que x et y sont non conjugués et achève la preuve de la proposition.

En fait, il est possible de retrouver, sur cet exemple, le résultat du théorème (2.9) par un calcul direct que nous indiquons rapidement.

Soit $x_t^\varepsilon(x)$ la diffusion de générateur $\varepsilon^2 L$ issue de x . L'équation (1.4) s'écrit ici :

$$\left\{\begin{array}{l} x_1^\varepsilon(t) = x_1 + \varepsilon w_t^1 \\ x_i^\varepsilon(t) = x_i + \varepsilon \displaystyle\int_0^t (\varepsilon w_s^1 + x_1)^{k_i} dw_s^2 , \qquad i \geq 2 \end{array}\right. \tag{2.19}$$

Conditionnellement à $w_\cdot^1$, $(x_i^\varepsilon(t))_{i \geq 2}$ est une variable gaussienne de moyenne $(x_i)_{i \geq 2}$, et de covariance $\varepsilon^2 \Gamma^\varepsilon(t)$ donnée par :

$$\Gamma_{ij}^\varepsilon(t) = \int_0^t (x_1 + \varepsilon w_s^1)^{k_i+k_j} ds , \qquad 2 \leq i,j \leq d \tag{2.20}$$

On obtient ainsi l'expression du noyau de la chaleur :

$$p_{\varepsilon^2}(x,y) = \frac{1}{(\sqrt{2\pi}\,\varepsilon)^d}\, e^{-\frac{|y_1-x_1|^2}{2\varepsilon^2}}\, E\left((\det \Gamma^\varepsilon(1))^{-\frac{1}{2}} \,\middle|\, \varepsilon w_1(1) = y_1 - x_1\right) \tag{2.21}$$

pour (x,y) tels que : $y_i - x_i = 0$ si $i \in \{2 \dots d\}$.

Or le pont brownien converge vers le segment $s \longrightarrow s(y_1 - x_1)$ lorsque ε tend vers zéro, et $\Gamma^{\varepsilon}(1)$ converge vers la matrice A introduite en (2.18).
Si $y_1 - x_1 \neq 0$, on peut montrer que $E((\det \Gamma^{\varepsilon}(1))^{-1/2} \mid \varepsilon w_1(1) = y_1 - x_1)$ converge vers $(\det A)^{-1/2}$ et on obtient ainsi l'équivalent :

$$p_{\varepsilon^2}(x,y) \sim \frac{(\det A)^{-1/2}}{\sqrt{2\pi}^{d}} \frac{1}{\varepsilon^{d}} e^{-\frac{|y_1-x_1|^2}{2\varepsilon^2}} ,$$

ce qui, compte tenu de (2.14), est exactement le résultat donné par la proposition (2.12) (ou le théorème (2.9)).
Le développement asymptotique complet s'obtiendrait en développant (en puissances de ε) $(\det \Gamma^{\varepsilon}(1))^{-1/2}$.
Ce calcul donne aussi la valeur des coefficients du développement au moins pour les points x,y tels que $x_i - y_i = 0 \quad 2 \leq i \leq d$ et $x_1 - y_1 \neq 0$.

III. NOYAU DE LA CHALEUR SUR LA DIAGONALE

1. Le théorème

Les résultats du II ne concernent pas le comportement asymptotique du noyau de la chaleur sur la diagonale, c'est-à-dire de $p_t(x,x)$ car comme on l'a vu (x,x) est dans le cut-locus si L n'est pas elliptique en x.
Nous allons donner ce comportement asymptotique dans le cas particulier où *le drift* X_0 *est nul*. Nous consacrerons une section à l'influence (surprenante) du drift sur ce comportement asymptotique.
L'étude faite ici est donc la première approche pour comprendre le comportement asymptotique du noyau de la chaleur sur le cut-locus.
Commençons par introduire quelques notations :
Soit, pour tout entier $k \geq 0$, l'espace vectoriel $C_k(x)$ engendré par les crochets des champs $(X_i)_{1\leq i\leq m}$, de longueur inférieure ou égale à k, pris au point $x \in \mathbb{R}^d$.
Par l'hypothèse (1.2), il existe un entier $r(x)$ tel que : $C_{r(x)}(x) = \mathbb{R}^d$. Notons $Q(x) = \sum_{k=1}^{r(x)} k(\dim C_k(x) - \dim C_{k-1}(x))$.
Soit la relation d'équivalence sur $\mathbb{R}^d$:

$$x \sim y \iff \forall k \in \mathbb{N} \quad \dim C_k(x) = \dim C_k(y) \tag{3.1}$$

et notons $[x]$ la classe d'équivalence de x.
Nous allons voir que le comportement de $p_t(x,x)$ dépend de ce que nous appellerons "la géométrie en x", i.e. la classe d'équivalence $[x]$.
On a en effet, avec les notations et les hypothèses précédentes (en particulier $X_0 = 0$) :

THÉORÈME (3.2) (cf. [4] et [14]).- *Pour tout* $n \in \mathbb{N}$, *on a le développement asymptotique à l'ordre* n

$$p_t(x,x) = \frac{1}{\sqrt{t}^{\,Q(x)}} \left(\sum_{k=0}^{n} c_k(x)\, t^k + t^{n+1} r_{n+1}(t,x) \right) \quad \text{où}$$

a) *les* c_k *sont des fonctions continues sur* $[x]$;

b) $c_0(x) > 0$;

c) *le développement est uniforme sur les compacts de* $[x]$: *si* K *est un compact inclus dans* $[x]$, *il existe* $t_0 > 0$ *tel que* :

$$\sup_{t\leq t_0} \sup_{y\in K} |r_{n+1}(t,y)| < \infty .$$

De plus, si $[x]$ *est d'intérieur non vide, les* c_k *sont des fonctions* C^∞ *sur tout ouvert inclus dans* $[x]$.

Remarque (3.3).- On peut aussi, en utilisant les résultats de [5], donner une expression explicite (mais très lourde), du premier coefficient $c_0(x)$ (voir [4]).

Nous allons, de nouveau, ne donner qu'une indication très rapide de la preuve du théorème (3.2) :

Comme pour celle du théorème (2.9), on commence par une inversion de Fourier :

$$p_{\varepsilon^2}(x,x) = \frac{1}{2\pi^d} \int E(e^{i\xi.(x_1^\varepsilon - x)})d\xi ,$$

où x^ε est donnée par (1.4). Il s'agit alors d'étudier l'annulation de $x_1^\varepsilon(x) - x$ lorsque ε tend vers zéro, en utilisant les développements de Taylor stochastiques introduits en [5]. Après avoir vérifié simplement que l'estimation de $p_{\varepsilon^2}(x,x)$ est un problème local, on montre, en utilisant la technique de relèvement de Rotschild et Stein sur un groupe nilpotent et les résultats de [5] sur les développements de Taylor stochastiques sur ces groupes qu'il existe une carte de $\mathbb{R}^d$ au voisinage de x telle que, dans cette carte :

$$x_1^\varepsilon(x) - x = \delta_\varepsilon Y^\varepsilon ,$$

où Y^ε est C^∞ en ε, où Y^o a une matrice de Malliavin inversible et où δ_ε est une dilatation de déterminant $\varepsilon^{Q(x)}$. On obtient ainsi, après le changement de variable $\zeta = \delta_\varepsilon \xi$:

$$p_{\varepsilon^2}(x,x) = \frac{1}{2\pi^d} \frac{1}{\varepsilon^{Q(x)}} \int d\zeta \, E(^{i\zeta.Y^\varepsilon}) .$$

En suivant l'idée de la preuve du théorème (2.9), on développe en puissance de ε, à ζ fixé, la quantité $E(e^{i\zeta.Y^\varepsilon})$. L'intégration par parties du calcul de Malliavin permet alors de montrer, du fait que Y_o a une matrice de Malliavin inversible, qu'il est possible d'intégrer en ζ ce développement asymptotique pour obtenir le théorème (3.2).

2. Exemples

a) *Le cas elliptique*

Dans le cas où L est elliptique en x, on a $Q(x) = d$ et les développements asymptotiques donnés par les théorèmes (2.9) et (3.2) coïncident puisqu'alors (x,x) n'est pas dans le cut-locus.

b) *Les groupes nilpotents*

Soit G un groupe de Lie nilpotent d'algèbre $\mathcal{G}$ avec :

$$\mathcal{G} = V_1 \oplus \ldots \oplus V_n \quad \text{ou} \quad [V_i,V_j] = V_{i+j} .$$

Si les $(X_i)_{1\le i\le m}$ sont des champs invariants à gauche sur G tels que les $(X_i(e))_{1\le i\le m}$ forment une base de V_1, et si $L = \frac{1}{2}\sum_1^m X_i^2$, alors il y a une unique classe d'équivalence pour la relation $\sim$ introduite en (3.1) et, pour tout $x \in G$, $Q(x) = \sum_1^m k \dim V_k$.

Le résultat du théorème (3.2) est alors évident par homogénéité et on a même $c_k \equiv 0 \quad \forall k > 1$.

Ceci est par exemple le cas du groupe d'Heisenberg.

c) *Reprise de l'exemple c) du II*

Reprenons l'opérateur introduit dans l'exemple c) du II.

Soit $x \in \mathbb{R}^d$ tel que $x_1 = 0$; on vérifie alors simplement que :

. si $k \notin \{1, k_1+1, \dots, k_d+1\}$, alors $C_k(x) = C_{k-1}(x)$;

. si $k = 1$, $\dim C_1(x) = 1$;

. si $k = k_i + 1$, $\dim C_{k_i+1}(x) = \dim C_{k_i}(x) + 1$.

D'où l'on tire que : $Q(x) = 1 + \sum_{j=2}^{d} (k_j + 1) = d + \sum_{j=2}^{d} k_j$.

Le résultat du théorème (3.2) peut de nouveau être retrouvé directement grâce à la formule (2.21) qui donne :

$$p_{\varepsilon^2}(x,x) = \frac{1}{\sqrt{2\pi}^d \varepsilon^d} E((\det \Gamma^\varepsilon(1))^{-1/2} \mid \varepsilon w_1(1) = 0) .$$

Or, puisque $x_1 = 0$, on a $\Gamma^\varepsilon_{ij}(1) = \varepsilon^{k_i+k_j} \int_0^1 w_1(1)^{k_i+k_j} ds$, d'où l'on tire :

$\det \Gamma^\varepsilon(1) = \varepsilon^{2\sum_2^d k_i} \det \Gamma^1(1)$, et donc :

$$p_{\varepsilon^2}(x,x) = \frac{1}{\varepsilon^{d+\sum_2^d k_i}} \frac{1}{\sqrt{2\pi}^d} E((\det \Gamma^1(1))^{-1/2} \mid w_1(1) = 0) ,$$

ce qui montre qu'ici on a : $C_k(x) = 0 \quad \forall k > 1$ et donne une expression explicite du terme dominant $c_0(x)$.

3. Volume des boules hypoelliptiques

Nous allons comparer ici le résultat du théorème (3.2) à l'estimation suivante due à Jerison et Sanchez-Calle [10] (voir aussi [12]) :

Avec les notations et les hypothèses précédentes (rappelons qu'ici $X_0 = 0$), on a :

$$\frac{K}{\mathrm{Vol}\, B(x,\sqrt{t})} e^{-C\frac{d^2(x,y)}{2t}} \le p_t(x,y) \le \frac{K'}{\mathrm{Vol}\, B(x,\sqrt{t})} e^{-C'\frac{d^2(x,y)}{2t}} , \tag{3.4}$$

où K , K' , C , C' sont des constantes positives strictement et où $\mathrm{Vol}\, B(x,\sqrt{t})$ désigne le volume de la boule hypoelliptique :

$$B(x,\sqrt{t}) = \{y \in \mathbb{R}^d / d(x,y) < \sqrt{t}\} .$$

Cette inégalité n'est précise, en temps petit, que sur la diagonale $y = x$:

$$\frac{K}{\mathrm{Vol}\, B(x,\sqrt{t})} \le p_t(x,x) \le \frac{K'}{\mathrm{Vol}\, B(x,\sqrt{t})} . \tag{3.5}$$

En utilisant le résultat du théorème (3.2), on obtient donc :

PROPOSITION (3.6).- *Pour tout* $x \in \mathbb{R}^d$ *et tout compact* L *inclus dans* $[x]$, *il existe un* $\rho_0 > 0$ *et des constantes* K , K' *telles que :* $\forall \rho \le \rho_0$, $\forall y \in L$:

$$\frac{K}{\rho^{Q(x)}} \le \mathrm{Vol}\, B(y,\rho) \le \frac{K'}{\rho^{Q(x)}} .$$

Ce résultat est une version assez rudimentaire du résultat de Nagel - Stein - Wainger [17] qui donne le comportement asymptotique précis du volume des boules hypoelliptiques lorsque le rayon tend vers zéro.
Réciproquement, en utilisant le résultat de Nagel - Stein - Wainger et le théorème (3.2), on retrouve l'encadrement (3.5) en temps petit.

4. Trace du semi-groupe de la chaleur

Nous posons ici le problème d'évaluer asymptotiquement une intégrale de la forme $\int_K p_t(x,x)\,dx$ où K est un compact de $\mathbb{R}^d$.

PROPOSITION (3.7).- *Si le compact* K *est inclus dans une classe d'équivalence* $[x_0]$ (i.e. "*la géométrie est constante sur* K "), *alors on a, pour tout* $n \geq 0$:

$$\int_K p_t(x,x)\,dx = \frac{1}{\sqrt{t}^{Q(x_0)}}\left(\sum_{k=0}^{n} t^k \int_K c_k(x,x)\,dx + O(t^{n+1})\right)$$

et $\int_K c_0(x,x)\,dx > 0$.

La proposition (3.7) est une conséquence immédiate du théorème (3.2) ; il suffit d'intégrer le développement de (3.2) qui est uniforme sur K .
Par contre, si K n'est pas inclus dans une unique classe d'équivalence (i.e. il y a bifurcation de la géométrie sur K), le théorème (3.2) ne permet pas en général de donner même un équivalent de $\int_K p_t(x,x)\,dx$.
Considérons l'exemple le plus simple d'une telle situation :
Soit, sur $\mathbb{R}^2$, l'opérateur de Grushin $L = \partial^2_{x_1} + x_1^2\partial^2_{x_2}$ (qui entre dans la classe traitée dans le c) des exemples).
Il est clair qu'il existe deux classes d'équivalence sur $\mathbb{R}^2$: l'axe $\Delta = \{x_1 = 0\}$ et son complémentaire. Sur Δ^c , L est elliptique et l'on a :

$$p_t(x,x) \sim \frac{c_0(x)}{t} \qquad \text{avec} \qquad c_0(x) = \frac{\text{cste}}{|x_1|} \tag{3.8}$$

Sur Δ , on a : $p_t(x,x) \sim \dfrac{c_0(x)}{t^{3/2}}$ avec $c_0(x) = \text{cste}$ (3.9)

Ainsi, si le compact K (que nous supposerons d'intérieur non vide) rencontre Δ , l'estimation de $\int_K p_t(x,x)\,dx$ n'est pas évidente.
Certes, dans ce cas, l'ensemble $K \cap \Delta$ est de mesure nulle ; néanmoins, on ne peut intégrer le développement donné par le théorème (3.2) sur $K \cap \Delta^c$, car le coefficient dominant $c_0(x)$ a un pôle en $x_1 = 0$, et $\int_K c_0(x)\,dx$ diverge.
Nous posons ici deux problèmes ouverts :

1) quel est, en général, le comportement des coefficients $c_k(x)$ du développement du noyau de la chaleur sur la diagonale, lorsque x tend, en restant dans la classe $[x_0]$, vers un point y du bord de $[x_0]$ où la géométrie est plus dégénérée que sur $[x_0]$, i.e. $Q(y) > Q(x_0)$?

On peut penser que, génériquement (au moins avec des coefficients analytiques), les

c_k ont une explosion du type $\frac{1}{\|x-y\|^{\nu}}$ où l'exposant ν est donné par la bifurcation de la géométrie en y ;

2) quel est le comportement asymptotique de $\int_K p_t(x,x)\,dx$ pour un compact K général ?

On peut conjecturer l'apparition de termes en $\frac{1}{t^{\alpha}(\log t)^{\beta}}$ dans ce développement asymptotique.

5. Influence du drift sur le comportement asymptotique de $p_t(x,x)$

Jusqu'ici, dans ce paragraphe, nous avons considéré le cas où X_0 est nul ; nous allons indiquer ici un phénomène surprenant si X_0 est non nul. Ce phénomène est étudié en détail dans un travail en collaboration avec R. Léandre ([7]).
Le résultat est que si, en x, le drift X_0 n'appartient pas à l'espace vectoriel $C_2(x)$ (engendré par les crochets de longueur inférieure ou égale à 2 des champs $(X_i)_{1\le i\le m}$), alors le noyau de la chaleur $p_t(x,x)$ (associé à $\frac{1}{2}\sum_1^m X_i^2 + X_0$), bien loin d'avoir une explosion en $\frac{1}{t^{Q(x)/2}}$ lorsque t tend vers zéro, a au contraire une décroissance exponentielle en $e^{-\frac{C(x)}{t^{\alpha(x)}}}$ avec, bien sûr, $\alpha(x) < 1$.
Une intuition assez vague de ce résultat peut être donnée comme suit :
Le drift X_0 agit sur la diffusion dans une échelle de temps t, le mouvement brownien n'agit, dans les directions des crochets de longueur ≥ 3 des champs X_i, que dans une échelle de temps inférieure à $t^{3/2}$, puisque dans ces directions la diffusion a pour coordonnées des intégrales itérées du brownien d'ordre ≥ 3 (voir [5]). Ainsi, le brownien "n'arrive plus à compenser l'action du drift" et la probabilité de retour en x dans l'échelle de temps t est exponentiellement faible.
Un exemple très simple illustre ce raisonnement fondé sur l'homogénéité :
Soit $N_{m,p}$ le groupe *libre* nilpotent à m générateurs $X_1 \dots X_m$ et de longueur p. $N_{m,p}$ (ou son algèbre) s'écrit donc :

$$N_{m,p} = V_1 \oplus \dots \oplus V_p ,$$

où V_i est engendré par les crochets de longueur i des $(X_j)_{1\le j\le m}$.
Si on considère un champ $X_0 \in V_k$, on vérifie simplement par homogénéité que, pour $x \in N_{m,p}$:

$$d\,(e^{tX_0}\,.\,x,x) = t^{\frac{2}{k}}\, d^2(e^{X_0}\,x,x) \tag{3.10}$$

où d désigne la distance hypoelliptique associée aux $(X_i)_{1\le i\le m}$.
Soit alors sur $N_{m,p}$ les opérateurs invariants à gauche :

$$L = \frac{1}{2}\sum_1^m X_i^2 \quad \text{et} \quad \bar{L} = \frac{1}{2}\sum_1^m X_i^2 + X_0 ,$$

où $X_0 \in V_p$, i.e. X_0 est dans le centre de $N_{m,p}$.

On a, puisque X_0 et L commutent :

$$\bar{p}_t(x,y) = p_t(e^{tX_0}x,y) \tag{3.11}$$

si $p_t(x,y)$ (respectivement $\bar{p}_t(x,y)$) désigne le noyau de la chaleur associé à L (respectivement à $\bar{L}$).

Par l'estimée (3.4) pour l'opérateur L , on a :

$$p_t(e^{tX_0}x,y) \le \frac{K}{\mathrm{Vol}\, B(x,\sqrt{t})} e^{-C\frac{d^2(e^{tX_0}x,y)}{t}} . \tag{3.12}$$

D'où, par (3.10), (3.11) et (3.12) :

$$\bar{p}_t(x,x) \le \frac{K}{\mathrm{Vol}\, B(x,\sqrt{t})} e^{-C\frac{d^2(e^{X_0}x,x)}{t^{(1-2/p)}}} , \tag{3.13}$$

ce qui montre que $\bar{p}_t(x,x)$ décroît exponentiellement <u>si $p \ge 3$</u> , et montre aussi que la minoration dans l'estimée (3.4) (ou (3.5)) est fausse si le drift X_0 n'est pas dans l'espace engendré par les crochets de longueur ≤ 2 des X_i .

Remarque (3.14).- Le cas où le drift X_0 est dans l'espace engendré par les crochets de longueur ≤ 2 des X_i est plus subtil. Si X_0 peut s'écrire :

$$X_0 = \sum_{i=1}^{m} f_i X_i + \sum_{i,j} f_{ij}[X_i,X_j] \tag{3.15}$$

où les fonctions f_i , f_{ij} sont C^∞ au voisinage de x , alors le théorème (3.2) reste valide. Par contre, si, pour tout x , $X_0(x) \in C_2(x)$ mais que l'on ne puisse écrire X_0 sous la forme (3.15), il peut se produire ou pas ce phénomène de décroissance exponentielle, comme le montre l'exemple où $X_1 = \partial_{x_1}$, $X_2 = x_1^n \partial_{x_2}$, $X_0 = x_1^p \partial_{x_2}$ sur $\mathbb{R}^2$.

Cet exemple, traité dans [7], montre que si $0 < p < n-1$ et si p est *pair*, alors, pour tout x , on a bien $X_0(x) \in C_2(x)$ et pourtant que le théorème (3.2) est faux (précisément $c_0(x) = 0$), pour les x tels que $x_1 = 0$. En ces points, on a décroissance exponentielle de $p_t(x,x)$.

BIBLIOGRAPHIE

[1] R. AZENCOTT - *Densité des diffusions en temps petit : développements asymptotiques*, Séminaire de Probabilités (1982-83), Lecture Notes in Maths 1059, p. 402-498, Springer-Verlag.

[2] R. AZENCOTT et al. - *Géodésiques et diffusions en temps petit*, Astérisque 84-85, SMF (1981).

[3] G. BEN AROUS - *Développement asymptotique du noyau de la chaleur hors du cut-locus*, à paraître aux Annales de l'E.N.S.

[4] G. BEN AROUS - *Développement asymptotique du noyau de la chaleur sur la diagonale*, à paraître aux Annales de l'Institut Fourier.

[5] G. BEN AROUS - *Flots et séries de Taylor stochastiques*, à paraître dans "Probability theory and related fields".

[6] G. BEN AROUS - *Méthodes de Laplace et de la phase stationnaire sur l'espace de Wiener.*

[7] G. BEN AROUS et R. LÉANDRE - En préparation.

[8] J.-M. BISMUT - *Large deviations and Malliavin calculus*, Progress in Maths n° 45, Birkhauser, Basel (1984).

[9] B. GAVEAU - *Principe de moindre action, propagation de la chaleur, estimées sous-elliptiques sur certains groupes nilpotents*, Acta Mathematica 139, p. 96-153 (1977).

[10] S. JERISON, A. SANCHEZ-CALLE - *Estimates for the heat kernel for a sum of squares of vector fields*, Indiana University Math. Journal, Vol. 35 #4 (86), p. 835-855.

[11] Y. KANNAÏ - *Off diagonal short time asymptotics for solution of diffusion equations*, CPDE 2(8), p. 751-830 (1977).

[12] S. KUSUOKA, D. STROOCK - *Applications of the Malliavin calculus*, Part III, à paraître.

[13] R. LÉANDRE - *Estimation en temps petit de la densité d'une diffusion dégénérée*, Note aux C.R.A.S., t. 301, Série 1, n° 17, p. 801 (1985).

[14] R. LÉANDRE - *Développement asymptotique de la densité de diffusions dégénérées*, à paraître au Journal of Functional Analysis.

[15] R. LÉANDRE - *Intégration dans la fibre*, à paraître dans "Probability theory and related fields".

[16] S.A. MOLCHANOV - *Diffusion processes and Riemannian geometry*, Russian Math. Survey 30, p. 1-53 (1975).

[17] A. NAGEL, E. STEIN, S. WAINGER - *Balls and metrics defined by vector fields I, Basic properties*, Acta Mathematica 155 (1985), p. 103-147.

[18] R. STRICHARTZ - *Sub-Riemannian geometry*, J. Diff. Geometry 24 (1986), p. 221-263, and a recent correction to this paper.

[19] S.R.S. VARADHAN - *Diffusion processes in a small time interval*, C.P.A.M. 20 (1967), p. 659-685.

ON TWO CLASSES OF SMOOTH MEASURES FOR SYMMETRIC MARKOV PROCESSES

Masatoshi FUKUSHIMA
Department of Mathematics. College of General Education.
Osaka University, Toyonaka, Osaka, 560 (Japan)

§1 Introduction

Given a right continuous strong Markov process $M = (\Omega, X_t, \zeta, P_x)$ on an appropriate topological space X, a *positive continuous additive functional* (PCAF in abbreviation) of M is by definition an extended real valued process $A_t(\omega)$, $t \geq 0$, $\omega \in \Omega$, adapted to an appropriate filtration such that, there exists a set $\mathcal{L} \subset \Omega$ with $P_x(\mathcal{L}) = 1$ for any $x \in X$ with the property that, for each $\omega \in \Omega$, $A_t(\omega)$ is as a function of t positive continuous, finite on $[0, \zeta(\omega))$, vanishing at 0 and additive in the sense that $A_{t+s}(\omega) = A_t(\omega) + A_s(\theta_t\omega)$ where θ_t denotes the (shift) operator from Ω to Ω satisfying $X_u(\theta_t\omega) = X_{t+u}(\omega)$, $u \geq 0$.

The set $\mathcal{L}$ is called the *defining set* of A. For instance, a bounded non-negative Borel function g on X produces a finite PCAF by

$$(1.1) \qquad A_t(\omega) = \int_0^{t\wedge\zeta(\omega)} g(X_s(\omega))ds$$

with its defining set being the full space Ω. Two PCAF's $A^{(1)}$ and $A^{(2)}$ are said to be *equivalent* if $P_x(A_t^{(1)} = A_t^{(2)},\ t \geq 0) = 1$ for any $x \in X$. We denote by A_{c1}^{+} the set of all PCAF's of M.

It is of fundamental importance in the theory of Markov process to give an analytical characterization of the family A_{c1}^{+}. To this end, we fix an M-excessive measure m on X and associate the measure $\mu = g\cdot m$ with the PCAF (1.1). μ is called the Revuz measure of (1.1). In general, a Borel measure μ on X is called the *Revuz measure* of $A \in A_{c1}^{+}$ with respect to m if

$$(1.2) \qquad \lim_{t\downarrow 0} \frac{1}{t} E_m\left(\int_0^t f(X_s)dA_a \right) = \langle f, \mu \rangle$$

for any non-negative Borel function f on X. Revuz [12] proved, under Hunt's strong duality assumption on M with respect to a reference measure m, that (1.2) gives a one-to-one correspondence between the family of the equivalence classes of A^{+}_{c1} and the class S_1 of Borel measures μ on X satisfying the follwoing properties: μ charges no semi-polar set and there exist Borel sets E_n increasing to X such that, for each n, $\mu(E_n)$ is finite, $I_{E_n}\cdot\mu$ is of bounded 1-potential and $P_x(\lim_{n\to\infty}\sigma_{X-E_n} \geq \zeta) = 1, x \in X$, σ_{X-E_n} being the hitting time of $X - E_n$. This result of Revuz essentially includes the previous ones of Mckean-Tanaka [10] and Wentzell-Dynkin[3] on PCAF's of multidimensional Brownian motions.

Notice that the above description of the class S_1 of measures is rather intricate and this is caused inevitably by the strigent finiteness requirement in the definiton of PCAF. For instance the functional $A_t = \int_0^t |X_s|^{-2}ds$ is not a PCAF of the d-dimensional Brownian motion because $P_0(A_t = \infty) = 1$ on account of the law of the iterated logarithm, while $|x|^{-2}dx$ is a nice Radon measure on R^d $(d \geq 3)$ charging no polar set. It is therefore natural to try to relax the finiteness requirement on PCAF's in some way or other to get a broader but simpler class of associated Revuz measures. It is also desirable to replace the strong duality assumption on M by weaker ones. For instance, when we study a symmetric Markov process in the Dirichlet space setting, we should require no a priori knowledge about the absolute continuity of the transition function with respect to the symmetrizing measure m.

In this connection, two meaningful generalizations of the concept of PCAF have appeared. One is the notion of the PCAF admitting exceptional polar set in the Dirichlet space setting. Another is the notion of the (continuous) homogeneous random measure (HRM in abbreviation) in the weak duality setting. Both notions still admit the associated Revuz measures by formula (1.2) and we denote by S (resp. $\tilde{S}$) the totality of the corresponding Revuz measures in the former (resp.latter) setting. A simple analytical description of S was given in [7] (see §2) which shows that S contains all positive Radon measures charging no set of zero capacity, while Getoor-Sharpe [9] proved that $\tilde{S}$ includes all σ-finite Borel measures charging no semi-polar set.

Speaking of the above mentioned example, the former setting regards the point set $\{0\}$ as exceptional for A_t and the latter regards A_t as a σ-finite positive measure on $(0,\infty)$ rather than on $[0,\infty)$. In both settings, $\mu = |x|^{-2}dx$ becomes the associated Revuz measure ($d \geq 2$). See C. Menendez [11] for S in the non-symmetric Dirichlet space setting and also Dynkin [4], Atkinson-Mitro [1] for earlier treatments of HRM's.

In this paper we look at things in a reverse direction. We start with the Dirichlet space setting and make use of the result in [7] about the construction of PCAF's admitting exceptional sets. We then impose an extra condition of the absolute continuity of the transition function and try to see how the class S_1 can be recovered in characterizing the family A^+_{c1} of PCAF's admitting no exceptional set.

Our description of S_1 replaces the condition of boundedness of potential in Revuz's by that of its finiteness together with finiteness of energy integral. The present procedure is very close to that of McKean-Tanaka [10] where the classical energy integral was used in the first step of the construction of the PCAF of the multidimensional Brownian motion. In this regards we mention an interesting paper by Dynkin [5] where HRM's were constructed for symmetric processes using a generalized notion of energy integrals.

Among PCAF and its stated two variants, the notion of HRM is perhaps most general, while the characterization of the two classes S_1 and S is indispensable in the study of transformations of Markov processes ([8]). Fitzsimmons has pointed out to the author (private communication) that, for a general Borel right process with a reference measure m, the same description of S_1 as Revuz (with an appropriate interpretation of bounded potentials) can be readily drawn from the Getoor-Sharpe characterization of $\tilde{S}$.

The author wishes to express his hearty thanks to Professors R.K. Getoor, M.J. Sharpe and P.J. Fitzsimmons for their truely valuable comments.

§2 Construction of a PCAF for $\mu \in S_0$ with a specific exceptional set

Let X be a locally compact separable metric space and m be a positive Radon measure on X with $\mathrm{supp}[m] = X$. We consider a regular Dirichlet form $(\mathcal{E},\mathcal{F})$ on $L^2(X;m)$ and a Hunt process $\mathbb{M} = (\Omega, X_t, \zeta, P_x)$ on X which is m-symmetric and associated with $(\mathcal{E},\mathcal{F})$.

A positive Radon measure μ on X is said to be *of finite energy integral* ($\mu \in S_0$ in notation) if

$$\int \varphi \, d\mu \le C \sqrt{\mathcal{E}_1(\varphi, \varphi)} \qquad \varphi \in \mathcal{F} \cap C_0(X)$$

for some constant $C > 0$. The 1-potential of μ is then well defined as an element $U_1\mu$ of $\mathcal{F}$ such that

$$\mathcal{E}_1(U_1\mu, \varphi) = \int \varphi \, d\mu \qquad \varphi \in \mathcal{F} \cap C_0(X).$$

A Borel measure μ on X is said to be *smooth* ($\mu \in S$ in notation) if μ charges no set of zero capacity and there exists an increasing sequence of compact sets $\{F_n\}$ such that $\mu(F_n) < \infty$ for each n, $\mu(X - \bigcup F_n) = 0$ and $\mathrm{Cap}(K - F_n) \longrightarrow 0$ for any compact set K. It is known that $\mu \in S$ if and only if there exists an increasing sequence of closed sets $\{F_n\}$ such that $I_{F_n} \cdot \mu \in S_0$ for each n, $\mu(X - \bigcup F_n) = 0$ and $P_x(\lim_{n\to\infty} \sigma_{X-F_n} \ge \zeta) = 1$ for q.e. Hence $S_0 \subset S$ and S contains any positive Radon measure charging no set of zero capacity.

Lemma 2.1 Denote by $\{p_t\}_{t>0}$ and $\{R_\alpha\}_{\alpha>0}$ the transition function and the resolvent of $\mathbb{M}$ respectively. The following conditions are equivalent :

(i) $p_t(x, \cdot) \in S$ for each $t > 0$ and $x \in X$.
(ii) $p_t(x, \cdot) \prec m$ for each $t > 0$ and $x \in X$.
(iii) $R_\alpha(x, \cdot) \in S$ for each $\alpha > 0$ and $x \in X$.
(iv) $R_\alpha(x, \cdot) \prec m$ for each $\alpha > 0$ and $x \in X$.

Here $\prec$ means the absolute continuity. Condition (i) is the same as saying that $p_t(x,\cdot)$ charges no set of zero capacity. Suppose $m(E) = 0$, then the symmetry yields $p_t I_E = 0$ m-a.e. and consequently q.e. because of the quasi-continuity of $p_t I_E$ ([7]). Therefore, if one assumes (i), then $p_{t+s} I_E(x) = \int p_t(x,dy) p_t I_E(y) = 0$, $x \in X$, getting (ii). (ii) and (iv) are known to be equivalent ([7]).

In what follows, we assume one of the equivalent conditions in Lemma 2.1. Then a set is of zero capacity if and only if it is M-polar (a Borel set B is said to be M-*polar* if $p_x(\sigma_B < \infty) = 0$, for any $x \in X$ where $\sigma_B = \inf \{ t > 0 : X_t \in B \}$). In particular any quasi-continuous function in $\mathcal{F}$ is finite except for an M-polar set.

A function $u \in L^2(X;m)$ is called *almost 1-excessive* if

$$u \geq 0, \quad e^{-t} T_t u \leq u \quad \text{m-a.e.}$$

where T_t is the L^2-realization of p_t. If u is almost 1-excessive, then $e^{-t} p_t u(x)$ increases as $t \downarrow 0$ for each $x \in X$. We denote the limitting function by $\tilde{u}$:

$$\tilde{u}(x) = \lim_{t \downarrow 0} p_t u(x) \ (\leq +\infty), \quad x \in X.$$

$\tilde{u}$ is a $\{p_t\}$-1-excessive function and $\tilde{u} = u$ m-a.e. If $u \in \mathcal{F}$ in addition, then $\tilde{u}$ is a quasi-continuous modification of u and accordingly the set $\{x \in X : \tilde{u}(x) = \infty \}$ is M-polar.

From now on, we use the term PCAF in the broader sense of [7] as compared with the traditional one. Thus PCAF is a functional A of M satisfying all the prperties stated at the beginning of § 1 except that we now impose on the defining set $\mathcal{L}$ of A a milder property that $p_x(\mathcal{L}) = 1$ for $x \in X - N$ for some set N of zero capacity. N is called the *exceptional set* of A. The totality of PCAF's of M is denoted by A_c^+. $A^{(1)}, A^{(2)} \in A_c^+$ are regarded to be equivalent if $P_x(A_t^{(1)} = A_t^{(2)},\ t \geq 0) = 1$ for q.e. $x \in X$. We have

$$A_{c1}^+ = \{ A \in A_c^+ : N = \phi \}$$

where A_{c1}^+ is the totality of PCAF's in the sense of the beginning of §1.

It has been proven in [7] that the equivalence classes of A_c^+ and the class S of smooth measures are in one-to-one correspondence by the Revuz formula (1.2). In particular, a PCAF A associated with a $\mu \in S_0$ was constructed in a way that the 1-potential of A is a quasi-continuous version of the 1-potnetial $U_1\mu$ of μ. Under the present absolute continuity assumption, the exceptional set of this A can be specified as follows:

Theorem 2.1 For $\mu \in S_0$, consider the M-polar set

$$N_\mu = \{ x \in X : U_1\mu^\sim(x) = +\infty \}$$

There exists then $A \in A_c^+$ with exceptional set N_μ such that

$$E_x\left(\int_0^\infty e^{-t} dA_t \right) = U_1\mu^\sim(x), \quad x \in X - N_\mu .$$

If $A^{(1)}$ and $A^{(2)}$ are as above, then

$$P_x(A_t^{(1)} = A_t^{(2)}, t \geq 0) = 1, \quad x \in X - N_\mu .$$

Proof by Theorem 5.1.1 of [7], there exists $A \in A_c^+$ with some exceptional set N and defining set $\mathcal{L}$ such that

$$u(x) = E_x\left(\int_0^\infty e^{-t} dA_t \right), \quad x \in X - N,$$

is a quasi-continuous version of $U_1\mu$. We may assume $P_\Delta(\mathcal{L}) = 1$ and $P_\Delta(\omega \in \Omega: A_t(\omega) = 0) = 1$. Let $A_t^\varepsilon(\omega) = A_{t-\varepsilon}(\theta_\varepsilon\omega)$, $\varepsilon > 0$, $\omega \in \Omega$, and, for a sequence $\varepsilon_n \downarrow 0$, $\mathcal{L}_0 = \bigcap \theta_{\varepsilon_n}^{-1} \mathcal{L}$.

Then $P_x(\mathcal{L}_0) = 1$ for any $x \in X$, because

$$P_x(\theta_{\varepsilon_n}^{-1}\mathcal{L}) = E_x(P_{X_{\varepsilon_n}}(\mathcal{L})) = \int_X p_t(x, dy) P_y(\mathcal{L}) + P_x(\zeta < \varepsilon)P_\Delta(\mathcal{L})$$

which is equal to 1 by the absolute continuity assumption.

If we set $A_t^n(\omega) = A^{\varepsilon_n}(\omega)$, $\omega \in \Omega$, then for each $\omega \in \mathcal{L}_0$

$$A_t^n(\omega) - A_t^m(\omega) = A_{\varepsilon_m - \varepsilon_n}(\theta_{\varepsilon_n}\omega), \quad m < n,$$

which means that $A_t^n(\omega)$ is increasing as $n \longrightarrow \infty$ locally uniformly in $t \in (0,\infty)$. On the other hand, from

$$A_t \leq e^t \int_0^t e^{-s} dA_s = e^t \int_0^\infty e^{-s} dA_s - e^t \int_t^\infty e^{-s} dA_s$$

holding on $\mathscr{L}$, we have

$$E_x(A_t) \le e^t\, u(x) - p_t u(x)\ , \quad x \in X - N,$$

and consequently,

$$E_x(A_t^{\varepsilon}) = E_x(A_{t-\varepsilon}(\theta_\varepsilon \omega)) = E_x(E_{X_\varepsilon}(A_{t-\varepsilon})) \le e^{t-\varepsilon} p_\varepsilon u(x) - p_t u(x)$$

for any $x \in X$.

Let, for $\omega \in \Omega$ and $t > 0$,

$$\tilde{A}_t(\omega) = \begin{cases} \lim_{n\to\infty} A_t(\omega) & \text{if the limit exists} \\ 0 & \text{otherwise.} \end{cases}$$

$\tilde{A}_t$ is then an adapted process. Moreover, for $x \in X - N_\mu$, we have from the preceding observations,

$E_x(\tilde{A}_t) \le \lim_{n\to\infty} E_x(A_t^n) \le e^{-t}\, \tilde{u}(x) - p_t \tilde{u}(x) < \infty$, $E_x(\tilde{A}_{0+}) = \lim_{t\downarrow 0} E_x(\tilde{A}_t)$
≤ 0

and hence

$$P_x(\ \tilde{A}_t < \infty,\ \tilde{A}_{0+} = 0\ ) = 1.$$

Therefore we can conclude that $\tilde{A}$ is a PCAF with exceptional set N_μ and with defining set

$$\mathscr{L} = \{\omega \in \Omega:\ \tilde{A}_t(\omega) < \infty \ \text{ for any } \ t > 0,\ \tilde{A}_{0+} = 0\ \}.$$

For $\omega \in \tilde{\mathscr{L}}$, we see that

$\int_{\varepsilon_n}^t e^{-s}\, dA_s^n = e^{-t} A_t^n + \int_{\varepsilon_n}^t e^{-s}\, A_s^n\, ds$ increases, as $n \longrightarrow \infty$, to $\int_0^t e^{-t}\, d\tilde{A}_s$. Hence, for $x \in X - N_\mu$,

$$E_x(\ \int_0^\infty e^{-s}\, d\tilde{A}_s) = \lim_{n\to\infty} E_x(\ \int_{\varepsilon_n}^\infty e^{-s}\, dA_s^n)$$

$$= \lim_{n\to\infty} E_x(e^{-\varepsilon_n} E_{X_{\varepsilon_n}}(\ \int_0^\infty e^{-s}\, dA_s)) = \lim_{n\to\infty} e^{-\varepsilon_n}\, p_{\varepsilon_n} u(x) = \tilde{u}(x).$$

To prove the uniqueness, take $A^{(1)}$, $A^{(2)}$ as in the first statement of Theorem 2.1. We know that $P_x(A_t^{(1)} = A_t^{(2)}, t > 0) = 1$ for q.e. $x \in X$. Hence, for any $x \in X - N_\mu$,

$$P_x(A_t^{(1)} - A_{\varepsilon_n}^{(1)} = A_t^{(2)} - A_{\varepsilon_n}^{(2)}) = E_x(\ P_{X_s}(A_{t-\varepsilon_n}^{(1)} = A_{t-\varepsilon_n}^{(2)})) = 1.$$

It now suffices to let $n \longrightarrow \infty$. q.e.d.

§3 A characterization of S_1 and a continuity property

We maintain the absolute continuity assumption embodied in Lemma 2.1. We introduce the class S_{01} of measures by

$$S_{01} = \{ \mu \in S_0 : U_1\mu^{\sim}(x) < \infty \quad \text{for any} \quad x \in X \}.$$

This class includes the family

$$S_{00} = \{ \mu \in S_0 : U_1\mu \quad \text{is m-essentially bounded and} \quad \mu(X) = 1 \}.$$

As was shown in [7], a set is of zero capacity if and only if it is μ-negligible for all $\mu \in S_{00}$. In this sense, the present class S_{01} is pretty large.

A Borel measure μ on X is said to be *smooth in the strict sense* if there exists a sequence of Borel quasi-closed sets $\{E_n\}$ increasing to X such that $I_{E_n} \cdot \mu \in S_{01}$ for each n and

$P_x(\lim_{n\to\infty} \sigma_{X-E_n} \geq \zeta) = 1$ for any $x \in X$. Here a set $E \subset X$ is called *quasi-closed* if, for any $\varepsilon > 0$, there exists an open set G with $\mathrm{Cap}(G) < \varepsilon$ such that $E - G$ is closed. The totality of the smooth measures in the strict sense is denoted by S_1. This description of the class S_1 is slightly different from Revuz's one stated in §1.

Recall the subfamily A^+_{c1} of A^+_c introduced in §2. Two members $A^{(1)}$, $A^{(2)}$ of A^+_{c1} are regarded to be equivalent if

$$P_x(A^{(1)}_t = A^{(2)}_t, \ t \geq 0) = 1, \quad \text{for any} \quad x \in X.$$

Theorem 3.1 The equivalence classes of A^+_{c1} are one-to-one correspondence with the class S_1 of smooth measures in the strict sense by the Revuz formula (1.2).

Proof For $\mu \in S_{01}$, there exists a unique $A \in A^+_{c1}$ (up to the equivalence) with Revuz measure μ by virtue of Theorem 2.1. This statement can be readily extended to $\mu \in S_1$. The converse assertion can be obtained just as in Revuz [12]. In fact, given $A \in A^+_{c1}$, it suffices to set $E_n = \{ x \in X : \varphi(x) \geq \frac{1}{n} \}$ with

$\varphi(x) = E_x(\int_0^\infty e^{-t} f(X_t) e^{-A_t} dt)$ which is quasi-continuous. q.e.d.

We note the inclusion $S_1 \subset S$. In fact, for $\mu \in S_1$ with $\{E_n\}$

in the statement of the definition, we can find a sequence of decreasing open sets G_n such that $\lim_{n\to\infty} \mathrm{Cap}(G_n) = 0$ and $F_n = E_n - G_n$ is closed for each n. Since $\sigma_{X-F_n} = \sigma_{X-E_n} \wedge \sigma_{G_n}$ and $P_x(\lim_{n\to\infty} \sigma_{G_n} = \infty) = 1$ q.e., we have $P_x(\lim_{n\to\infty} \sigma_{X-F_n} < \zeta) = 0$ q.e., proving that $\mu \in S$.

For $\mu \in S_1$, the corresponding element of A_{c1}^{+} is denoted by $A(\mu)$. Let us study a continuity of this correspondence.

Theorem 3.2 (i) For $\mu_1, \mu_2 \in S_1$ and $a, b \geq 0$,

$$A(a\mu_1 + b\mu_2) = aA(\mu_1) + bA(\mu_2).$$

(ii) For $\mu_1, \mu_2 \in S_1$ with $\mu_1 \geq \mu_2$, $A(\mu_1) \geq A(\mu_2)$.

(iii) If $\mu_n, \mu \in S_1$ and $\mu_n \uparrow \mu$, then $A(\mu_n) \uparrow A(\mu)$, namely,

$$P_x(\, A_t(\mu_n) \uparrow A_t(\mu),\ n \to \infty,\ t \geq 0\,) = 1,\ x \in X.$$

Proof (iii) Suppose first that $\mu_n, \mu \in S_{01}$ and $\mu_n \uparrow \mu$. Then we have $U_1\mu_n^{\sim}(x) \uparrow U_1\mu^{\sim}(x)$, $n \to \infty$, $x \in X$, because the increasing limit u of $U_1\mu_n^{\sim}$ is easily seen to be the $\mathcal{E}_1$-limit of $U_1\mu_n$ as well and, from $\mathcal{E}_1(U_1\mu_n, \varphi) = \langle \varphi, \mu_n \rangle$, $\varphi \in \mathcal{F} \cap C_0$, follows $\mathcal{E}_1(u, \varphi) = \langle \varphi, \mu \rangle$, $\varphi \in \mathcal{F} \cap C_0$. On the other hand, $A^n = A(\mu_n)$ is increasing by (ii) and $A_t^{n,1} = \int_0^t e^{-t}\, dA_s^n$ is increasing in n as well. Let $B_t = \lim_{n\to\infty} A_t^n$, $B_t^1 = \lim_{n\to\infty} A_t^{n,1}$. We have $B_t \leq A_t = A_t(\mu)$ and $B_t^1 \leq A_t^1 = \int_0^t e^{-s}\, dA_s$. Since $E_x(A_\infty^{n,1}) = U_1\mu_n^{\sim}(x)$, we get $E_x(B_\infty^1) = U_1\mu^{\sim}(x) = E_x(A_\infty^1)$, $x \in X$, and consequently $0 \leq B_t^1 - A_t^1 \leq B_\infty^1 - A_\infty^1 = 0$ and $B_t = A_t$ as was to be proved.

Next take any $\mu_n, \mu \in S_1$ with $\mu_n \uparrow \mu$. Let $\{E_\ell\}_{\ell=1}^{\infty}$ be a sequence of sets corresponding to μ. Since $I_{E_\ell} \cdot \mu \in S_{01}$ implies $I_{E_\ell} \cdot \mu_n \in S_{01}$, $\{E_\ell\}_{\ell=1}^{\infty}$ is a sequence well corresponding to μ_n for

each n. But, for each fixed ℓ, $I_{E_\ell}\cdot\mu_\mu \uparrow I_{E_\ell}\cdot\mu$ as $n \to\infty$, and hence we have from the preceding observation
$A_t(I_{E_\ell}\cdot\mu_n) \uparrow A_t(I_{E_\ell}\cdot\mu)$, $t \geq 0$, P_x-a.e. for every $x \in X$.
Consequently $A_t(\mu_n) \uparrow A_t(\mu)$ P_x-a.s. on $t < \sigma_{X-E_\ell}$ for each ℓ and the proof is complete because $P_x(\lim_{\ell\to\infty} \sigma_{X-E_\ell} \geq \zeta) = 1$, $x \in X$.
q.e.d.

Consider, as an example, the Brownian motion on R^d with $d \geq 2$, which is symmetric with respect to the Lebesgue measure and satisfies the absolute continuity condition. The measure $\mu(dx) = |x|^{-2}dx$ belongs to S. However $\mu \notin S_1$. Otherwise, the PCAF $A_t^n = \int_0^t I_{\{|X_s|\geq \frac{1}{n}\}}\cdot|X_s|^{-2}ds$ corresponding to $\mu_n(dx) = I_{\{|x|\geq\frac{1}{n}\}}\cdot|x|^{-2}\,dx$ should increase to an element of A_{c1}^+ by virtue of Theorem 3.2, a contradiction to the observation made in §1.

References

[1] B. Atkinson and J. Mitro, Applications of Revuz and Palm measures for additive functionals in weak duality, Seminar on stochastic processes 1982, 23-50, Birkhäuser.

[2] C. Dellacherie et P.A. Meyer, Probabilités et potentiel, Ch. XII á XVI, Hermann, Paris, 1987.

[3] E.B. Dynkin, Markov processes, Springer Verlag, 1965.

[4] E.B. Dynkin, Additive functionals of Markov processes and stochastic systems, Ann.Inst. Fourier(Grenoble) 25(1975), 177-200.

[5] E.B. Dynkin, Green's and Dirichlet spaces associated with fine Markov processes, J. Funct.Anal.47(1982),381-418.

[6] P.J. Fitzsimmons and R.K. Getoor, Revuz measures and time changes, to appear.

[7] M. Fukushima, Dirichlet forms and Markov processes, Kodansha and North Holland, 1980.

[8] M. Fukushima and Y. Oshima, On skew product of symmetric

diffusion processes, to appear.

[9] R.K. Getoor and M.J. Sharpe, Naturality , standardness and weak duality for Markov processes, Z. Wahrscheinlichkeitstheorie verw. Gebiete, 67(1984), 1-62.

[10] H.P. McKean and H. Tanaka, Additive functionals of the Brownian path, Memoire Coll.Sci. Univ. Kyoto, A. Math.,33(1961), 479-506.

[11] S.C. Menendez, Fonctionnelles additives à un ensemble polaire près to appear in Stochastics

[12] D. Revuz, Measures associeés aux fonctionnelles additives de Markov I, Trans. Amer. Math. 70(1959), 43-72.

THE HYDRODYNAMICAL LIMIT FOR SCALAR GINZBURG-LANDAU MODEL ON R

Tadahisa FUNAKI
Department of Mathematics. Faculty of Science.
Nagoya University, Nagoya, 464 (Japan)

1. Introduction.

It is one of the most important problems in the theory of non-equilibrium statistical mechanics to derive macroscopic evolution equations like hydrodynamical equations (compressible Euler equation etc.) from underlying microscopic dynamics. Several attempts have been made to analyze the hydrodynamical behavior of some stochastic models in the last years. De Masi, Ianiro, Pellegrinotti and Presutti [1] is a review paper devoted to these problems.

In this paper we discuss one-dimensional time-dependent Ginzburg-Landau model of conservative type. This model has been investigated for the study of dynamic critical phenomena (see Hohenberg and Halperin [5]). The following stochastic partial differential equation (SPDE) is introduced to describe the model:

$$(1.1)\qquad dS_t(x) = -\Delta^2 S_t(x)dt + \Delta\{U'(S_t(x))\}dt + \sqrt{2}\,\nabla dw_t(x),$$
$$t > 0,\ x \in R;\ \Delta = \frac{d^2}{dx^2},\ \nabla = \frac{d}{dx},$$

where $w_t(x)$ is a cylindrical Brownian motion on the space $L^2(R,dx)$ (see Section 2). The self-potential U is a real-valued function on R and satisfies the following technical condition:

$$U(s) = \frac{\gamma}{2} s^2 + V(s),\quad \gamma > 0,\quad V \in C_b^3(R).$$

The solution $S_t(x) \in R$ of the SPDE (1.1) represents the random time evolution of order parameters such as magnetization being distributed on the space R.

The object of this paper is to know the macroscopic behavior of the Ginzburg-Landau model. It is accomplished by investigating the asymptotic property of the scaled process $S_t^\varepsilon(x) = S_{t/\varepsilon^2}(x/\varepsilon)$ as the parameter ε tends to 0. We shall derive a nonlinear partial differential equation (PDE) by proving a law of large numbers

(Theorem 2). See [3] for some results on a non-scalar Ginzburg-Landau model, i.e. the case where $S_t(x)$ takes values in a manifold.

2. Main results.

2.1. Existence and uniqueness theorem for the SPDE (1.1). We introduce a family of real Hilbert spaces $\mathbf{H}_r = L^2(R, e^{-r|x|}dx)$, $r \in R$, having norms defined by

$$|S|_r = \{\int_R S^2(x)e^{-r|x|}dx\}^{1/2} \quad , \quad S \in \mathbf{H}_r \ .$$

Let $\mathbf{H}_e = \bigcap_{r>0}\mathbf{H}_r$ and $\mathbf{H}_e^* = \bigcup_{r>0}\mathbf{H}_{-r}$ be a countably Hilbertian space and its dual, respectively. We sometimes consider a weak topology $\sigma(\mathbf{H}_e, \mathbf{H}_e^*)$ on the space $\mathbf{H}_e$. With this topology it is denoted by $\mathbf{H}_{e,w}$. Let $\mathcal{C}_r$, $r \in R$, be the space of all $S \in \mathcal{C} \equiv C(R)$ satisfying

$$|||S|||_r = \sup_{x\in R} |S(x)|e^{-r|x|} < \infty.$$

The space $\mathcal{C}_e = \bigcap_{r>0} \mathcal{C}_r$ is a countably normed space.

The mathematical meaning of the SPDE (1.1) can be given by re-writing it formally into a stochastic integral equation:

$$(2.1) \qquad S_t(x) = \int_R q(t,x,y)S_0(y)dy - \sqrt{2}\int_0^t\int_R q_y(t-u,x,y)dw_u(y)dy$$
$$+ \int_0^t\int_R q_{yy}(t-u,x,y)V'(S_u(y))dudy \quad , \quad t \geq 0, \ x \in R,$$

where $q(t,x,y)$ is a fundamental solution of a parabolic operator $\frac{\partial}{\partial t} + \Delta^2 - \gamma\Delta$. The subscripts to q mean its derivatives with respect to those variables, e.g., $q_y = \partial q/\partial y$. The initial data S_0 is always taken from the space $\mathbf{H}_e$. We assume that the cylindrical Brownian motion w_t is defined on a probability space $(\Omega, \mathcal{F}, P)$ with reference family $\{\mathcal{F}_t\}$. Namely w_t is an $\{\mathcal{F}_t\}$-adapted $\mathcal{S}'(R)$-valued continuous process satisfying $w_0 = 0$ and

$$E[\exp\{\sqrt{-1} \langle w_t - w_s, \phi\rangle\} | \mathcal{F}_s] = \exp\{-(t-s)\|\phi\|^2_{L^2(R,dx)}/2\} \quad \text{a.s.},$$

for every $\phi \in \mathcal{S}(R)$ and $t \geq s \geq 0$.

The stochastic process $S_t = \{S_t(x,\omega); x \in R\}$, $t \geq 0$, is called a solution of the SPDE (1.1) if it is $\{\mathcal{F}_t\}$-adapted, jointly measurable in (t,x,ω) and satisfies the stochastic integral equation (2.1) with probability one. We state the existence and uniqueness theorem for the SPDE (1.1) without proof.

Theorem 1 (i) There exists a solution S_t of the SPDE (1.1). Every solution satisfies that $S_t \in C((0,\infty), \mathcal{C}_e)$ with probability one.
(ii) Let S_t and S_t' be two solutions of the SPDE (1.1). If $S_0 = S_0'$, then we have $S_t = S_t'$, $t \geq 0$, with probability one.
(iii) Suppose $S_0 \in \mathcal{C}_e$, then $S_t \in C([0,\infty), \mathcal{C}_e)$ with probability one.

2.2. Hydrodynamical limit. Now we introduce the hydrodynamical scaling:

(2.2) $$t \longrightarrow t/\varepsilon^2 , \quad x \longrightarrow x/\varepsilon , \quad \varepsilon > 0,$$

for the solution of the SPDE (1.1). Then the scaled process $S_t^\varepsilon(x) = S_{t/\varepsilon^2}(x/\varepsilon)$ is equivalent in law to the solution of the following SPDE:

$(1.1)_\varepsilon$ $$dS_t^\varepsilon(x) = -\varepsilon^2\Delta^2 S_t^\varepsilon(x)dt + \Delta\{U'(S_t^\varepsilon(x))\}dt + \sqrt{2\varepsilon}\,\nabla\, dw_t(x),$$
$$t > 0, \quad x \in R.$$

Set $\langle S,\phi\rangle = \int_R S(x)\phi(x)dx$ for $S \in \mathcal{C}$ and $\phi \in C_0^\infty(R)$. Our main theorem is formulated as follows.

Theorem 2 Suppose $\|V''\|_\infty = \sup_{s\in R}|V''(s)|$ is sufficiently small. If the initial data $S_0^\varepsilon \equiv S_0$ of $(1.1)_\varepsilon$ is independent of ε and belongs to the class $C_b^3(R)$, then $\langle S_t^\varepsilon,\phi\rangle$ converges to a non-random function $\langle \rho_t,\phi\rangle$ in probability as ε tends to 0 for every $t > 0$ and $\phi \in C_0^\infty(R)$. Here $\rho_t = \rho_t(x)$ is a classical solution of the nonlinear PDE:

(2.3) $$\frac{\partial}{\partial t}\rho_t(x) = \nabla\{d(\rho_t(x))\nabla\rho_t(x)\} \quad , \quad t > 0 \ , \quad x \in R,$$

with an initial data S_0.

The diffusion coefficient $d(\rho)$ appearing in (2.3) is positive and determined in the following manner: Let Ω_λ, $\lambda \in R$, be the ground state of a self-adjoint operator

$$H_\lambda = -\frac{1}{2}\frac{d^2}{ds^2} + \{U(s)-\lambda s\} \quad ,$$

which is defined on the space $L^2(R,ds)$. Namely Ω_λ is a positive and normalized eigenfunction of H_λ corresponding to its minimal eigenvalue. Define a function $\bar{\rho}$ by

$$\bar{\rho}(\lambda) = \int s\Omega_\lambda^2(s)ds \quad , \quad \lambda \in R.$$

Then $\bar{\rho}$ is real analytic and strictly increasing in λ. It can be shown that $\lim_{\lambda\to\pm\infty}\bar{\rho}(\lambda) = \pm\infty$. The function $d(\rho)$ is a derivative

$$d(\rho) = \bar{\lambda}'(\rho)$$

of an inverse function $\bar{\lambda} = \bar{\lambda}(\rho)$ of $\bar{\rho} = \bar{\rho}(\lambda)$.

The smoothness of the coefficient $d(\rho)$ combined with the condition $S_0 \in C_b^3(R)$ of the initial data guarantees the existence and uniqueness of classical solutions of the PDE (2.3). See Ladyzenskaya, Solonnikov and Ural'ceva [6].

3. Heuristic argument.

Before giving the outline of the proof of Theorem 2, we explain how a physical argument leads us to the conclusion of Theorem 2. Therefore the nature of this section is quite heuristic, but must be helpful to understand the general feature of the problem of the hydrodynamical limit.

The physicists prefer the following form (3.1) to the SPDE (1.1):

(3.1) $$dS_t(x) = - A\{D\mathcal{H}(x,S_t)\}dt + \sqrt{2A}\, dw_t(x) \quad , \quad A = -\Delta ,$$

where $D\mathcal{H}(x,S)$ denotes the functional (Fréchet) derivative of the Ginzburg-Landau-Wilson free energy:

$$\mathcal{H}(S) = \int_R \{\tfrac{1}{2}|\nabla S(x)|^2+U(S(x))\}\, dx .$$

The rewritten form (3.1) suggests that the equation (1.1) has an invariant measure μ formally defined by

$$d\mu(S) = e^{-\mathcal{H}(S)} \text{"}dS\text{"} / \text{normalization}, \quad \text{"}dS\text{"} = \Pi_{x\in R}\, dS(x).$$

The role of the operator A is that it causes the equation (3.1) to have the total spin (or density) $\int S(x)dx$ as a (formal) conserved quantity. Therefore the equation (1.1) might carry as its invariant measures not only μ but also a one-parameter family $\{\mu_\lambda\}_{\lambda\in R}$ of measures given by

$$d\mu_\lambda(S) = \exp\{-\mathcal{H}(S)+\lambda\int S(x)dx\} \text{"}dS\text{"} / \text{normalization}, \ \lambda \in R.$$

The parameter λ represents the strength of the external field.

What we can prove mathematically is the following assertion: Define the probability measure μ_λ on the space $\mathcal{C}$ by a weak limit

$$d\mu_\lambda(S) = \lim_{\ell\to\infty} Z_{\ell,\lambda}^{-1}\exp[-\int_{-\ell}^{\ell} \{U(S(x))-\lambda S(x)\}dx]\, d\mu_{-\ell,\ell}^{0,0}(S) ,$$

where $Z_{\ell,\lambda}$ is a normalizing constant and $\mu_{-\ell,\ell}^{0,0}$ is a probability distribution on $\mathcal{C}$ of the pinned Brownian motion $\{S(x);\ x \in R\}$ satisfying $S(x) = 0$ for $|x| \geq \ell$. Then μ_λ is an invariant measure of the SPDE (1.1). It is known that considering x to be a time variable, under the distribution μ_λ, $S(x)$ is a symmetric diffusion process with reversible measure $d\nu_\lambda(s) \equiv \Omega_\lambda^2(s)ds$; especially, $\mu_\lambda\{S(x)\varepsilon ds\} = d\nu_\lambda(s)$ for every $x \varepsilon R$.

A quick derivation of the PDE (2.3) is now possible by assuming the so-called principle of hydrodynamics (the local equilibrium approximations): There exists a function $\lambda(t,x)$ such that for each $(t,x) \varepsilon (0,\infty)\times R$ the distribution on R of the random variable $S_t^\varepsilon(x)$ converges weakly to the probability measure $\nu_{\lambda(t,x)}$ as ε tends to 0. This assumption seems true, since the hydrodynamical scaling makes the system evolve so rapidly that $S_t^\varepsilon(x)$ is likely to converge weakly to one of the 1-dimensional distributions $\{\nu_\lambda\}_{\lambda\varepsilon R}$ of equilibrium states $\{\mu_\lambda\}_{\lambda\varepsilon R}$ of the SPDE (1.1). Let us consider only the asymptotic behavior of the mean $\rho_t^\varepsilon(x) = E[S_t^\varepsilon(x)]$ for simplicity. Then from the equation $(1.1)_\varepsilon$ one might have

$$\frac{\partial}{\partial t}\rho_t^\varepsilon(x) = \Delta E[U'(S_t^\varepsilon(x))] + o(\varepsilon) , \quad \varepsilon \downarrow 0.$$

The principle of hydrodynamics implies

$$\lim_{\varepsilon\downarrow 0} E[U'(S_t^\varepsilon(x))] = \int U'(s)d\nu_{\lambda(t,x)}(s) .$$

However, using integration by parts, we see that the right hand side is equal to $\lambda(t,x)$. Therefore, if the limit $\rho_t(x) = \lim_{\varepsilon\downarrow 0} \rho_t^\varepsilon(x)$ exists, then ρ_t solves the equation (2.3) because $\lambda(t,x) = \bar{\lambda}(\rho_t(x))$.

4. Outline of the proof of Theorem 2.

The actual proof of Theorem 2 is not given as in the manner explained in Section 3. It seems not so easy to establish the principle of hydrodynamics for our model. We shall follow and extend the method due to Fritz [2], in which a discrete version of the Ginzburg-Landau model was discussed. The proof will be divided into three main steps. See [4] for more detailed description.

(a) In the first step we investigate the spatial scaling limit, which will be formulated as the law of large numbers, for special but sufficiently wide class of initial distributions. Let Λ be the family of all C^2-functions $\lambda(\cdot)$ on R such that the derivatives $\lambda'(\cdot)$ have compact supports. We can associate with each $\lambda(\cdot) \varepsilon \Lambda$ a probability measure $\mu_{\lambda(\cdot)}$ on $\mathcal{C}$ by taking a weak limit:

$$d\mu_{\lambda(\cdot)}(S) = \lim_{\ell\to\infty} Z_{\ell,\lambda(\cdot)}^{-1}\exp[-\int_{-\ell}^{\ell}\{U(S(x))-\lambda(x)S(x)\}dx]\, d\mu_{-\ell,\ell}^{0,0}(S),$$

where $Z_{\ell,\lambda(\cdot)}$ is a normalizing constant. The function $\lambda(\cdot)$ exhibits a profile of the strength of the spatially dependent external field.

Let τ_ε and σ_ε be two mappings defined by $(\tau_\varepsilon\lambda)(x) = \lambda(\varepsilon x)$ respectively $(\sigma_\varepsilon S)(x) = S(x/\varepsilon)$ for $x \in R$, $\lambda(\cdot) \in \Lambda$ and $S \in \mathcal{C}$. Consider an image measure $\mu_{\lambda(\cdot),\varepsilon} \equiv \mu_{\tau_\varepsilon\lambda(\cdot)}\circ\sigma_\varepsilon^{-1}$, $0 < \varepsilon < 1$, of $\mu_{\tau_\varepsilon\lambda(\cdot)}$ under the mapping σ_ε. This transformation corresponds to the spatial scaling "$x \longrightarrow x/\varepsilon$" introduced in (2.2), which is acting on $\{\mu_{\lambda(\cdot)}\}$ in such a manner that the profile $\lambda(\cdot)$ is kept to be unchanged at each position. We can prove the following assertion on an asymptotic behavior of $\{\mu_{\lambda(\cdot),\varepsilon}\}$:

<u>Proposition 1</u> The probability measure $\mu_{\lambda(\cdot),\varepsilon}$ converges weakly to the δ-distribution $\delta_{\overline{\rho}(\lambda(\cdot))}$ on the space $\mathbf{H}_{e,w}$ as ε tends to 0.

In order to prove this proposition, we see the convergence of every finite dimensional marginal distribution of $\mu_{\lambda(\cdot),\varepsilon}$. Indeed, as ε becomes close to 0, the random variables $\{S(x_i)\}_{1\le i\le n}$ under the probability distribution $\mu_{\lambda(\cdot),\varepsilon}$ for distinct $\{x_i\}$'s in R are asymptotically independent and we may regard $\lambda(\cdot)$ to be almost constant around every fixed position $x \in R$. Remembering that $\mu_\lambda\circ S(0)^{-1} = \nu_\lambda$ if $\lambda \equiv$ constant, we can actually prove

<u>Lemma 1</u> For every $n = 1,2,\cdots$, $x_1 < x_2 < \cdots < x_n$ and $\xi_i \in C_b(R)$, $1 \le i \le n$, we have

$$\lim_{\varepsilon\downarrow 0} E^{\mu_{\lambda(\cdot),\varepsilon}}\left[\prod_{i=1}^{n} \xi_i(S(x_i))\right] = \prod_{i=1}^{n} \int_R \xi_i d\nu_{\lambda(x_i)} \quad .$$

Let $\mathcal{D}$ be the class of all functions Ψ defined on the space $\mathcal{C}$ having the form:

$$(4.1) \qquad \Psi(S) = \psi(\langle S,\phi_1\rangle,\cdots,\langle S,\phi_k\rangle) \quad , \quad S \in \mathcal{C},$$

with $k = 1,2,\cdots$, $\psi = \psi(\alpha_1,\cdots,\alpha_k) \in C_0^\infty(R^k)$ and $\phi_1,\cdots,\phi_k \in C_0^\infty(R)$. A consequence of Lemma 1 is that $E^{\mu_{\lambda(\cdot),\varepsilon}}[\Psi]$ converges to $\Psi(\overline{\rho}(\lambda(\cdot)))$ as ε tends to 0 for every $\Psi \in \mathcal{D}$. Therefore the proof of Proposition 1 can be completed if the relative compactness of the family $\{\mu_{\lambda(\cdot),\varepsilon}\}_{0<\varepsilon<1}$ of probability measures is shown. To this end, we use Prokhorov's criterion; it is true even on the space $\mathbf{H}_{e,w}$, since this space is completely regular and its every compact subset is metrizable. Noting that $B(\mathbf{b}) \equiv \{S \in \mathbf{H}_e \ ; \ |S|_r \le b_r,\ r > 0\}$ is compact in the space $\mathbf{H}_{e,w}$, we may only prove that for every $\delta > 0$

$$(4.2) \qquad \inf_{0<\varepsilon<1} \mu_{\lambda(\cdot),\varepsilon}\{B(\mathbf{b})\} > 1-\delta$$

holds with some positive sequence $\mathbf{b} = \{b_r\}_{r>0}$.

The following lemma is a conclusion of the so-called FKG inequality. We put $\lambda_+ = \sup_{x\in R}\lambda(x)$ and $\lambda_- = \inf_{x\in R}\lambda(x)$.

Lemma 2 $\quad \mu_{\lambda_-,\varepsilon} \leq \mu_{\lambda(\cdot),\varepsilon} \leq \mu_{\lambda_+,\varepsilon}$

In this lemma, $\mu_1 \leq \mu_2$ for two probability measures μ_1 and μ_2 on $\mathcal{C}$ means that we can construct, on a proper probability space, $\mathcal{C}$-valued two random variables S_1 and S_2 distributed by μ_1 and μ_2, respectively, in such a way that $S_1 \leq S_2$ (i.e., $S_1(x) \leq S_2(x)$, $x \in R$) holds with probability one. The estimate (4.2) now follows from a uniform estimate

$$\sup_{x\in R,\ 0<\varepsilon<1} \int S^2(x)d\mu_{\lambda(\cdot),\varepsilon}(S) < \infty,$$

which is an easy consequence of Lemma 2.

(b) Define a (formal) generator of the $\mathcal{C}$-valued process S_t^ε, the solution of the SPDE $(1.1)_\varepsilon$, by

$$\mathcal{G}^\varepsilon\Psi(S) = \varepsilon \sum_{i,j=1}^{k} \frac{\partial^2\psi}{\partial\alpha_i\partial\alpha_j}(<S,\phi_1>,\dots,<S,\phi_k>)<-\Delta\phi_i,\phi_j>$$

$$+ \sum_{i=1}^{k} \frac{\partial\psi}{\partial\alpha_i}(<S,\phi_1>,\dots,<S,\phi_k>)\{-\varepsilon^2<S,\Delta^2\phi_i> + <U'(S(\cdot)),\Delta\phi_i>\},$$

for $\Psi \in \mathcal{D}$ having the form (4.1). Set $T_t^\varepsilon\Psi(S) = E_S[\Psi(S_t^\varepsilon)]$, $S \in \mathcal{C}$, $\Psi \in \mathcal{D}$, where S is an initial data of the SPDE $(1.1)_\varepsilon$. The second step is devoted to deriving the following formula.

Proposition 2 For every $\lambda(\cdot) \in \Lambda$ and $\Psi \in \mathcal{D}$, we have

$$(4.3) \qquad \int T_t^\varepsilon \mathcal{G}^\varepsilon \Psi d\mu_{\lambda(\cdot),\varepsilon} = \int <\Delta\lambda(\cdot),DT_t^\varepsilon\Psi(\cdot,S)>d\mu_{\lambda(\cdot),\varepsilon},$$

and therefore

$$(4.4) \qquad \frac{\partial}{\partial t}\int T_t^\varepsilon\Psi\, d\mu_{\lambda(\cdot),\varepsilon} = \int <\Delta\lambda(\cdot),DT_t^\varepsilon\Psi(\cdot,S)>d\mu_{\lambda(\cdot),\varepsilon},$$

where D represents the Fréchet derivative.

The proof of this proposition is completed by showing two types of approximation theorems for S_t^ε. In the first place we approximate S_t^ε by a solution $S_t^{\varepsilon,\ell}(x)$, $x \in [-\ell,\ell]$, $\ell \geq 1$, of a finite volume SPDE; that is, an SPDE of the form $(1.1)_\varepsilon$ which is restricted on a finite interval $[-\ell,\ell]$ putting proper boundary conditions at both edges $\pm\ell$. Secondly, the Galerkin method can be used as usual to approximate $S_t^{\varepsilon,\ell}$ further by a finite dimensional process. Namely, we consider a formal Fourier series expansion of the solution of the finite volume

SPDE based on the family of eigenfunctions of the operator $-\Delta^2+\gamma\Delta$ defined on the interval $[-\ell,\ell]$. Then, for this finite dimensional process, we can easily establish a formula similar to (4.3). It is nothing but a formula of integration by parts. Taking the limit we finally arrive at the equality (4.3).

(c) We are involved in this step with proving the relative compactness of two families $\{T_t^\varepsilon\Psi\}_{0<\varepsilon<1}$ and $\{DT_t^\varepsilon\Psi\}_{0<\varepsilon<1}$ in suitable spaces, respectively, for every $\Psi \in \mathcal{D}$. The assumption of $\|V''\|_\infty$ being sufficiently small is required. We omit the detail.

We can, then, take the limit of ε tending to 0 in the equality (4.4). The operator D is closed in a proper sense and therefore, from Proposition 1, every limit Ψ_t of $\{T_t^\varepsilon\Psi\}_{0<\varepsilon<1}$ satisfies an equation:

$$\frac{\partial}{\partial t}\Psi_t(\bar{\rho}(\lambda)) = \langle\Delta\lambda(\cdot),D\Psi_t(\cdot,\bar{\rho}(\lambda))\rangle$$

for every $\lambda \in \Lambda$. Hence we obtain

$$\frac{\partial}{\partial t}\Psi_t(\rho) = \langle\Delta\{\dddot{\bar{\lambda}}(\rho)\},D\Psi_t(\cdot,\rho)\rangle$$

for every $\rho \in \Lambda$. However this is just the Liouville equation associated with the nonlinear PDE (2.3) so that the solution Ψ_t is given by

$$\Psi_t(\rho) = \Psi(\rho_t(\cdot,\rho))$$

for every $\rho \in C_b^3(R)$. Here $\rho_t(\cdot,\rho)$ is the unique classical solution of the PDE (2.3) with initial data ρ. These observations are concluded into

$$\lim_{\varepsilon\downarrow 0} E_S[\Psi(S_t^\varepsilon)] = \Psi(\rho_t(\cdot,S)) \quad , \quad S \in C_b^3(R),$$

and this completes the proof of Theorem 2.

5. Remark.

A similar method works also for the Ginzburg-Landau model of non-conservative type. The SPDE which describes the model is

$$(5.1) \qquad dS_t(x) = \Delta S_t(x)dt + U'(S_t(x))dt + \sqrt{2}dw_t(x) \quad , \quad t > 0,\ x \in R.$$

We have taken A = identity in (3.1).

An appropriate scaling for this equation is

$$(5.2) \qquad t \longrightarrow t \quad , \quad x \longrightarrow x/\varepsilon \quad , \quad \varepsilon > 0.$$

Note that no scaling is considered for the time variable. This is because the SPDE (5.1) has no conserved quantity. The limiting hydrodynamical equation is an ordinary differential equation:

$$(5.3) \qquad \frac{\partial}{\partial t} \rho_t(x) = - \bar{\lambda}(\rho_t(x)) \ , \quad t > 0,$$

for each $x \in R$.

The solution ρ_t of (5.3) has a property:

$$\lim_{t\to\infty} \rho_t(x) = \rho_* \ , \quad x \in R,$$

where $\rho_* \in R$ is a unique point such that $\bar{\lambda}(\rho_*) = 0$. This contrasts highly with the conservative case where the limiting equation (2.3) has a family of stationary solutions, that is, constant functions.

REFERENCES

[1] A. De Masi, N. Ianiro, S.Pellegrinotti and E. Presutti, A survey of the hydrodynamical behavior of many-particle systems, in Non-equilibrium Phenomena II, edited by J.L. Lebowitz and E.W. Montroll, North-Holland, Amsterdam, 1984.

[2] J. Fritz, On the hydrodynamical limit of a one-dimensional Ginzburg-Landau lattice model. The a priori bounds, J. Statis. Phys., 47 (1987), 551-572.

[3] T. Funaki, On diffusive motion of closed curves, in Probability Theory and Mathematical Statistics, edited by S. Watanabe and Yu.V. Prokhorov, Lecture Notes in Mathematics, Springer, to appear.

[4] T. Funaki, Derivation of the hydrodynamical equation for one-dimensional Ginzburg-Landau model, IMA Preprint Series, #328, University of Minnesota, 1987.

[5] P.C. Hohenberg and B.I. Halperin, Theory of dynamic critical phenomena, Rev. Mod. Phys., 49 (1977), 435-479.

[6] O.A. Ladyzenskaya, V.A. Solonnikov and N.N. Ural'ceva, Linear and quasilinear equations of parabolic type, AMS, Translations of mathematical monographs, 23, 1968.

SHORT TIME ASYMPTOTICS FOR FUNDAMENTAL SOLUTIONS OF DIFFUSION EQUATIONS

Nobuyuki IKEDA
Department of Mathematics. Faculty of Science
Osaka University, Toyonaka, Osaka, 560 (Japan)

Shigeo KUSUOKA
Research Institute for Mathematical Sciences
Kyoto University, Kyoto, 606 (Japan)

1. Introduction

Let D be the exterior of a *strictly convex* domain with smooth boundary ∂D in the Euclidean space R^n with endowed the standard flat metric g. We consider the heat equation with the Neumann boundary condition:

$$(1.1)\qquad \begin{aligned} \frac{\partial u}{\partial t} &= \frac{1}{2}\Delta u & (t,x) &\in (0,\infty)\times D \\ \frac{\partial u}{\partial n} &= 0 & (t,x) &\in (0,\infty)\times \partial D \end{aligned}$$

where Δ is the Laplacian on R^n and $\partial/\partial n$ denotes the normal derivative of u. Then there exists a continuous positive minimal fundamental solution $p(t,x,y)$, $(t,x,y) \in (0,\infty) \times \bar{D} \times \bar{D}$, of (1.1). The purpose of the present paper is to study the asymptotic behavior of $p(t,x,y)$ as $t\downarrow 0$.

We first fix two points x and y in $\bar{D}$, $(x \neq y)$, such that there exists a unique shortest curve (minimal geodesic) γ in $\bar{D}$ from $x = \gamma(0)$ to $y = \gamma(1)$. Then Varadhan's result ([13]) implies that as $t\downarrow 0$

$$(1.2)\qquad \log p(t,x,y) \sim -\rho(x,y)^2/(2t)$$

where $\rho(x,y)$ is the length of the curve γ. Our problem is to show an improved form of the asymptotic formula (1.2) which reflects the shape of the boundary ∂D near the curve γ. It is easy to show by using the "principle of not feeling the boundary" that if γ lines completely within D, $p(t,x,y)$ has the similar asymptotic behavior to that of the fundamental solution of the heat equation in the whole space R^n, (cf. Molchanov [12]). Hence, from now on, we will restrict ourselves to the case when $\Gamma \equiv \partial D \cap \{\gamma(t);\ 0 \leq t \leq 1\} \neq \phi$. As stated in [5], by using the continum product integral, Buslaev [3] derived non-rigorously the following elegant improvement of (1.2): If the length of Γ is positive,

$$(1.3) \qquad \log p(t,x,y) = -\rho(x,y)^2/(2t) - C/(2t)^{1/3} + o(t^{-1/3}), \text{ as } t\downarrow 0$$

where C is a positive constant which depends on the curvature of the boundary ∂D. When D is the outor domain of a ball in R^n, one of the authors first gave a rigorous justification of (1.3), ([5]). In this case, the problem can be reformulated in terms of asymptotic behavior of integrals related to the reflecting Brownian motion on $\overline{D}$ constructed by skew product of typical diffusion processes. As we first announced in the lecture of the Warwick symposium on SDE and Appl., 1984/85, the formula (1.3) for a general domain D mentioned above will be also shown under several mild assumptions. By using the theory of stochastic differential equations we first formulate the problem in terms of integrals related to the Brownian sheet. Applying the Cameron-Martin formula and the generalized Malliavin calculus based on the Brownian sheet ([2], [7], [8]), we finally arrive at the evaluation of the asymptotics of a Wiener integral

$$(1.4) \qquad E[\exp\{-z\int_0^1 k(t)\,|w(t)|\,dt\}] \qquad \text{as} \quad z\downarrow 0$$

where $E[\cdot]$ denotes the expectation with respect to the probability law of the standard Brownian bridge. Here $k(t)$, $0 \leq t \leq 1$, is a non-negative function such that the value $k(t)$ at t depends on the curvature of ∂D at $\gamma(t)$.

It should be mentioned that in [4] Hsu also independently gave a probabilistic proof of results related to (1.3) based on a method initiated by Molchanov [12], inspiring by [5]. He has described the fine structure of the right hand side of (1.3) in the case when $x,y \in \partial D$. By using a different approach from the proof mentioned below he also reduce the problem to the evaluation of the integral (1.4).

As stated in Buslaev [3], the problem discussed here is closely related to the asymptotic behavior of diffractions on smooth convex surfaces. For recent works on the theory of diffractions, see the article [10] by Melrose and Taylor together with the bibliography. It should also be remarked that the asymptotics of solutions of diffusion equations can be reduced to the asymptotic behavior of solutions of hyperbolic equations by the Hadamard formula, (cf. Kannai [6]).

In Section 2 we describe the fine structure of the constant C in (1.3). An outline of the proof of the main result is given in Section 4. Although the details here will be omitted, the sketch of the proof given in Section 4 will make points clear. Some results related to the problem discussed here and details of the proof will be published elsewhere.

2. Assumptions and main result

As stated in Section 1, we fix two points x and y in $\bar{D}$ which satisfy the following:

Assumption 2.1. There exists a unique shortest curve γ in $\bar{D}$ from x to y and $t_* < t^*$ where

$$t_* = \inf\{t;\ \gamma(t) \in \partial D\} \quad \text{and} \quad t^* = \sup\{t;\ \gamma(t) \in \partial D\}.$$

We call this curve γ the minimal geodesic from x to y, ([5]). Without loss of generality, we take the following parametrization:

$$|\dot{\gamma}(t)| = \rho(x,y), \qquad 0 \leq t \leq 1$$

where $\dot{\gamma}(t)$ denotes the velocity vector of $\gamma(t)$. Let $\mathcal{H}$ be the space defined by $\mathcal{H} = \{h\colon [0,1] \longrightarrow R^n$; each component h^i of $h = (h^1, h^2, \cdots, h^n)$ is absolutely continuous function on $[0,1]$ with square integrable derivative$\}$. For each $h \in \mathcal{H}$, we define the action integral of h by

$$\mathcal{E}[h] = \frac{1}{2}\int_0^1 |\dot{h}(t)|^2 dt, \quad \dot{h}(t) = (\dot{h}^1(t), \dot{h}^2(t), \cdots, \dot{h}^n(t))$$

where $\dot{h}^i(t) = dh^i(t)/dt$, $i = 1,2,\cdots,n$. We also set

$$H(x,y) = \{h \in \mathcal{H};\ h(t) \in \bar{D},\ 0 \leq t \leq 1,\ h(0) = x \ \text{and}\ h(1) = y\},$$

$$K(x,y) = \{k;\ k : [0,1] \longrightarrow \mathcal{H} \ \text{is a}\ C^{(2)}\text{-mapping and}\ k_s(0) = x,\ k_s(1) = y \ \text{for}\ s \in [0,1]\}.$$

As in smooth Riemannian manifolds, the action integral $\mathcal{E} : H(x,y) \longrightarrow R$

takes on its minimum precisely at the minimal geodesic γ, under Assumption 2.1. For the proof in case of smooth Riemannian manifold, see [11].

<u>Lemma</u> 2.1. Let $k \in K(x,y)$. If $k_0(t) = \gamma(t)$, $0 \leq t \leq 1$, then

$$(2.1)\quad (\frac{d}{ds}\,\&[k_s])\Big|_{s=0} = \int_{t_*}^{t^*} <(\frac{dk_s}{ds})\Big|_{s=0}(t),\ n(\gamma(t))>N(t)\,dt$$

where n is the inner ward unit normal vector field on ∂D and $N(t)$, $t_* \leq t \leq t^*$, is the function defined by

$$(2.2)\quad N(t) = \alpha_{\gamma(t)}(\dot{\gamma}(t),\ \dot{\gamma}(t)),\qquad t \in [t_*,\ t^*]$$

where α denotes the second fundamental form on ∂D.

<u>The sketch of the proof of Lemma 2.1</u>. Let $(\xi^1,\xi^2,\cdots,\xi^{n-1})$ be a system of local coordinates in a region of the boundary ∂D. Then the inclusion mapping from ∂D to R^n determines n smooth functions

$$x^1(\xi),x^2(\xi),\cdots,x^n(\xi),\qquad \xi = (\xi^1,\xi^2,\cdots,\xi^{n-1}).$$

Let $\xi(t) = (\xi^1(t),\xi^2(t),\cdots,\xi^{n-1}(t))$ be the coordinate of $\gamma(t)$ with respect to the coordinate system above. Then we have

$$(2.3)\quad \ddot{\gamma}^k(t) = \sum_{i,j=1}^{n-1} \frac{\partial^2 x^k}{\partial\xi^i\partial\xi^j}(\xi(t))\dot{\xi}^i(t)\dot{\xi}^j(t) + \sum_{i=1}^{n-1}\frac{\partial x^k}{\partial\xi^i}(\xi(t))\ddot{\xi}^i(t)$$
$$k = 1,2,\cdots,n\ .$$

Hence

$$\sum_{k=1}^{n} \frac{\partial x^k}{\partial\xi^\ell}(\xi(t))\ddot{\gamma}^k(t) = 0\ ,\qquad \ell = 1,2,\cdots,n-1$$

which implies

$$\ddot{\gamma}^k(t) \perp T_{\gamma(t)}(\partial D)\qquad k = 1,2,\cdots,n\ .$$

Combining this with (2.3) and the definition of the second fundamental form α, we have

$$\ddot{\gamma}(t) = -\alpha_{\gamma(t)}(\dot{\gamma}(t),\dot{\gamma}(t))n(\gamma(t))\qquad t \in [t_*,t^*]\ .$$

Since

$$(\frac{d}{ds}\,\mathcal{E}\,[k_s])\,|_{s=0} = -\int_{t_*}^{t^*} <(\frac{dk_s}{ds})\,|_{s=0}(t),\ddot{\gamma}(t)>dt,$$

this completes the proof of (2.1).

Keeping Lemma 2.1 in mind, we introduce a space of variations of the minimal geodesic γ as follows: Let $V(x,y)$ be the space of elements k of $K(x,y)$ such that

(i) $k_0 = \gamma$, (ii) $(\frac{d}{ds}k_s)\,|_{s=0} \not\equiv 0$, (iii) $k_s(t) \in \bar{D}$, $(s,t) \in [0,1]\times[t_*,t^*]$,

(iv) $<(\frac{d}{ds}k_s)\,|_{s=0}(t),\ n(\gamma(t))> = 0, \quad t \in [t_*,t^*]$.

<u>Definition</u> 2.1. x and y are called *non-conjugate along* γ if for every $k \in V(x,y)$,

$$(\frac{d^2}{ds^2}\,\mathcal{E}[k_s])\,|_{s=0} > 0 .$$

<u>Remark</u>. It is easy to see that for $k \in V(x,y)$

$$(\frac{d^2}{ds^2}\,\mathcal{E}[k_s])\,|_{s=0}$$
$$= 2\,\mathcal{E}\,[(\frac{d}{ds}k_s)\,|_{s=0}] + \int_{t_*}^{t^*} <(\frac{d^2}{ds^2}k_s)\,|_{s=0}(t),n(\gamma(t))>N(t)\,dt .$$

We now are in a position to state our main result.

<u>Theorem</u>. If x and y are non-conjugate along γ, then, as $t\downarrow 0$,

$$-\log p(t,x,y)$$
$$(2.4)\qquad = \rho(x,y)^2/(2t) + \lambda_1\int_{t_*}^{t^*} |N(s)|^{2/3}ds/(2t)^{1/3} + o(t^{-1/3})$$

where λ_1 is the first eigenvalue of

$$(2.5)\qquad \begin{aligned} &\frac{d^2}{d\xi^2}u - \xi u = -\lambda u\,, \quad \xi \in (0,\infty) \\ &u^+(0) = 0\,, \end{aligned}$$

and $N(t)$, $t_* \leqq t \leqq t^*$ is the function given by (2.2).

Roughly speaking, (2.4) means that the *major contribution* to the short time asymptotics of p comes from the critical point (minimal geodesic) of action integral $\mathcal{E}$ and the *deviation* from the major contribution depends on the one-side (restricted) first variation of action integral $\mathcal{E}$ at the critical point γ which is described by the integral of the second fundamental form of ∂D along γ.

3. Reformulation of the problem

Let us consider the following mapping:

$$(3.1) \qquad \Phi : [0,\infty) \times \partial D \longrightarrow \bar{D}$$

given by

$$(3.2) \qquad \Phi(a^0,a) = a + a^0 n(a) \qquad \text{for} \quad (a^0,a) \in [0,\infty) \times \partial D.$$

Then Φ is a diffeomorphism from $[0,\infty) \times \partial D$ onto $\bar{D}$. We again write g for $(\Phi^{-1})_*(g)$, i.e. we again denote by g the Riemannian metric induced by Φ^{-1} from the standard flat metric on R^n. Now it is clear that g can be naturally extended to a Riemannian metric on $R \times \partial D$ as follows:

$$g_{(a^0,a)} = g_{(|a^0|,a)} \qquad \text{for} \quad (a^0,a) \in R \times \partial D .$$

Hence we have a realization of the symmetric double $(R \times \partial D, g)$ of the Riemannian manifold $(\bar{D}, g)$. Since the second fundamental form on ∂D is strictly positive, the Riemannian metric $g_{(a^0,a)}$ on $R \times \partial D$ given above is not smooth and it is only Lipschitz continuous as a^0 crosses the hypersurface $N = \{(0,a)\,;\ a \in \partial D\}$. However the heat equation

$$(3.3) \qquad \frac{\partial u}{\partial t} = \frac{1}{2}Lu \qquad \text{on} \quad R \times \partial D$$

still has the continuous positive minimal fundamental solution $q(t,\xi,\eta)$, $(t,\xi,\eta) \in (0,\infty) \times (R \times \partial D) \times (R \times \partial D)$, where L is the Laplace-Beltrami operator with respect to g, (cf. Aronson [1]). Then, by the Kelvin reflection principle, we have

$$p(t,\ \Phi(a^0,a),\ \Phi(b^0,b))$$
$$= q(t,\ (a^0,a),\ (b^0,b)) + q(t,\ (-a^0,a),\ (b^0,b))$$
$$\text{for}\quad (a^0,a),\ (b^0,b) \in [0,\infty) \times \partial D\ .$$

Hence our concern is in the asymptotic behavior of $q(t,\ (a^0,a),\ (b^0,b))$ as $t\downarrow 0$.

4. Sketch of the proof of Theorem

The purpose of this section is to explain the strategy how one can prove (2.4). We first note that

(4.1) $$\Phi_*(\frac{\partial}{\partial a^0}) \perp \Phi_*(T_a(\partial D)) \quad\text{and}\quad |\Phi_*(\frac{\partial}{\partial a^0})| = 1$$
$$\text{for}\quad (a^0,a) \in [0,\infty) \times \partial D\ .$$

Hence the restriction of the operator L to $[0,\infty) \times \partial D$ can be written in the form

(4.2) $$L = \frac{1}{2}(\frac{\partial}{\partial a^0})^2 + \frac{1}{2}\sum_{i=1}^{d} V_i^2 + V_0 + b\frac{\partial}{\partial a^0}$$
$$\text{on}\quad [0,\infty) \times \partial D,$$

where V_i, $i = 0,1,\cdots,d$ are smooth vector fields on $[0,\infty) \times \partial D$ such that

$$V_i(a^0,a) \in T_a(\partial D) \quad\text{for}\quad (a^0,a) \in [0,\infty) \times \partial D,\ i = 0,1,\cdots,d$$

and b is a smooth function on $[0,\infty) \times \partial D$. Furthermore we may assume, without loss of generality, by Varadhan's result that the coefficients of V_i, $i = 0,1,\cdots,d$, b and their derivatives of all order are bounded. Moreover, for later use, we extended these to smooth functions on $R \times \partial D$ whose derivatives of all orders ($\geqq 0$) are bounded, i.e.

$$V_i \in C_b^\infty(R \times \partial D \longrightarrow T(\partial D)),\quad i = 0,1,\cdots,d$$

and

$$b \in C_b^\infty(R \times \partial D)\ .$$

From now on, we fix the following notation

$$\underline{u} = (u^0, u) = \Phi^{-1}(x) \quad \text{and} \quad \underline{v} = (v^0, v) = \Phi^{-1}(y).$$

Let $C^\alpha = C^\alpha([0,1] \longrightarrow R)$, $0 < \alpha < 1/2$, be the space of α-Hölder continuous functions on $[0,1]$ and $\{W^s(t) = W(s,t);\ 0 \leq s,\ t \leq 1\}$ be the d-dimensional Brownian sheet. We also consider a one-dimensional Brownian bridge $\{w^0(t),\ 0 \leq t \leq 1\}$ independent of $\{W^s(t),\ 0 \leq t \leq 1\}$. We denote by μ and P^s the probability laws of $\{w^0(t),\ 0 \leq t \leq 1\}$ and $\{W^s(t),\ 0 \leq t \leq 1\}$ respectively. For $\psi \in C^\alpha$, let $\{X(t;\ s,\ W^s;\ \psi)\}$ be the solution of the following stochastic differential equation on ∂D

$$(4.3) \qquad \begin{aligned} dX(t) &= \sum_{i=1}^{d} V_i(\psi(t), X(t)) \circ dW^{s,i}(t) + sV_0(\psi(t), X(t))dt \\ X(0) &= u \end{aligned}$$

where $W^s(t) = (W^{s,1}(t), W^{s,2}(t), \cdots, W^{s,d}(t))$. For simplicity, we write $X^{s,\psi}(t)$ or $X^{s,\psi}(t, W^s)$ for $X(t;\ s, W^s;\ \psi)$.

We consider Hilbert spaces H and H^0 defined by

$$H = \{h\colon [0,1] \longrightarrow R^d;\ h(0) = 0, \text{ each component is absolutely continuous and } \|h\|_H^2 = \int_0^1 |\dot{h}(t)|^2 dt < \infty\}$$

and

$$H^0 = \{h\colon [0,1] \longrightarrow R;\ \text{absolutely continuous},\ h(0) = h(1) = 0 \text{ and } \|h\|_{H^0}^2 = \int_0^1 |\dot{h}(t)|^2 dt < \infty\}.$$

We now define the basic space of $\mathscr{A}(\mathscr{A};\partial D)$ with respect to the generalized Malliavin calculus based on the Brownian sheet $\{W^s(t)\}$ by the same way as that in Kusuoka [7], Section 2. Then we obtain a nice mapping $X^{s,\cdot}(1, W^s)\colon C^\alpha \ni \psi \longrightarrow X^{s,\psi}(1, W^s) \in \mathscr{A}(\mathscr{A};\partial D)$. For details, see Kusuoka [7], Sections 2 and 4. Next, setting

$$\ell(t;\ u_0,\ v_0) = (1 - t)u_0 + tv_0, \quad 0 \leq t \leq 1,$$

we define a mapping $\Phi_1\colon C^\alpha \longrightarrow C^\alpha$ by

$$\Phi_1(w)(t) = |\ell(t;\ u_0,v_0) + w(t)|, \qquad 0 \leq t \leq 1.$$

Then $\tilde{q}(s,\underline{u},\underline{v}) = p(s,\Phi(\underline{u}),\Phi(\underline{v}))$ can be expressed in the following form:

$$\begin{aligned}(4.4)\qquad &\tilde{q}(s,\underline{u},\underline{v}) \\ &= Ks^{-\frac{1}{2}}\exp[-|u_0 - v_0|^2/2s]E^{\mu\otimes P^s}[\delta_v(X(1,s,W^s;\ \Phi_1(\sqrt{s}w^0)) \\ &\qquad \times M^{(s)}(W^s,\ \sqrt{s}w^0)]\end{aligned}$$

where

$$M^{(s)}(W^s,\ \sqrt{s}w^0) = \exp[\sqrt{s}\int_0^1 b(\Phi_1(\sqrt{s}w^0(t)),W^s(t))dw^0(t) - \frac{s}{2}\int_0^1 |b(\Phi_1(\sqrt{s}w^0(t)),W^s(t))|^2 dt] ,$$

K is the positive constant independent of s and δ_v is the Dirac delta at v. Here $E^{\mu\otimes P^s}[\cdot]$ denotes the expectation with respect to the product probability $\mu\otimes P^s$ of μ and P^s.

To compute the asymptotic behavior of the right hand side of (4.4) by using the ideas in Bismut [2] and Kusuoka [7], we consider the following equation;

$$(4.5)\qquad \begin{cases} \frac{d}{dt}X(t) = \sum_{i=1}^{d} V_i(\psi(t),X(t))\dot{h}^i(t) \\ \qquad\qquad\qquad\qquad\qquad\qquad , \quad h \in H \text{ and } \psi \in C^\alpha. \\ X(0) = u \end{cases}$$

Let $X^{0,\psi}(t;\ h)$ be the solution of (4.5) and let

$$\Psi = \{\psi : [0,1] \longrightarrow [0,\infty);\ \text{absolutely continuous and } \int_0^1 |\dot{\psi}(t)|^2 dt < \infty\} .$$

Setting, for $(\psi,h) \in \Psi \times H$,

$$\phi(t;\ \psi,h) = \Phi(\psi(t),X^{0,\psi}(t,h)), \qquad 0 \leq t \leq 1$$

and using (4.1), we obtain that if $\|\psi\|_\infty$ $(= \sup_{0\leq t\leq 1} |\psi(t)|)$ is sufficiently small, then

$$\mathcal{E}[\phi(\cdot;\ \psi,h)]$$
$$= \frac{1}{2}\{\int_0^1 |\dot{\psi}(t)|^2 dt + \int_0^1 |\Phi_*(\sum_{i=1}^d \dot{h}^i(t) V_i(\psi(t), X^{0,\psi}(t,h))|^2 dt\} .$$

Note that

$$\int_0^1 |\dot{\psi}(t)|^2 dt = (\psi(1) - \psi(0))^2 + \|\psi(\cdot) - \ell(\cdot;\ \psi(0),\psi(1))\|_{H^0}^2 .$$

Therefore there exists a unique $(\bar{h}^0,\bar{h}) \in H^0 \otimes H$ such that

$$\|\bar{h}^0\|_{H^0}^2 + \|\bar{h}\|_H^2 = \inf\{\|h^0\|_{H^0}^2 + \|h\|_H^2;\ (h^0,h) \in H^0 \otimes H,$$
$$v = X(1;\ 0,h;\ \Phi_1(h^0))\} .$$

Moreover, we have

$$\mathcal{E}[\phi(\cdot;\ \Phi_1(\bar{h}^0),\bar{h})] = \frac{1}{2}\{|u_0 - v_0|^2 + \|\bar{h}^0\|_{H^0}^2 + \|\bar{h}\|_H^2\}$$
$$= \inf\{\mathcal{E}[\phi(\cdot;\ \psi,h)];\ (\psi,h) \in \Psi \times H,\ \psi(0) = u_0, \psi(1) = v_0,$$
$$v = X(1;\ 0,h;\ \psi)\} .$$

Then we have $\gamma(t) = \phi(t;\ \Phi_1(\bar{h}^0),\bar{h})$, $0 \leq t \leq 1$. By applying the Cameron-Martin formula to (4.4), we obtain

$$\tilde{q}(s,\underline{u},\underline{v}) = Ks^{-\frac{1}{2}}\exp[-\mathcal{E}[\gamma]/s]$$
$$\times E^{\mu\otimes P^s}[\exp[-\frac{1}{\sqrt{s}}(\bar{h}^0,w^0)_{H^0} - \frac{1}{s}(\bar{h},W^s)_H]$$
$$\times \delta_v(X(1;\ s,W^s + \bar{h};\ \Phi_1(\sqrt{s}w^0 + \bar{h}^0)))M^{(s)}(W^s + \bar{h},\sqrt{s}w^0 + \bar{h}^0)] .$$

We set

$$\Phi_2(w^0)(t) = |\ell(t;\ u_0,v_0) + \bar{h}^0(t) + w^0(t)|$$
$$F(s,W^s,\psi) = X(1;\ s,\bar{h} + W^s;\ \psi) \qquad \psi \in \Psi.$$

Now a differential operator D of order one and the mapping $DF(0,0,\Phi_2(0))$ are defined as in Kusuoka [7], Section 2 (also see Bismut [2]). Then $\dim DF(0,0,\Phi_2(0))(H) = n-1$. Setting $K = \mathrm{Im}\, DF(0,0,\Phi_2(0))^*$, where $*$ means the adjoint, we split H to $H_1 \oplus K$ where $H_1 = K^{\perp}$. Then $\bar{h} \in H_1$. Corresponding to this decomposition, we have a pseudo orthogonal decomposition of W^s:

$$W^s = W_1^s + W_K^s$$

where $W_K^s = P_K W^s$ and P_K denotes the orthogonal projection in H onto K. We denote by P_1^s the probability law of $\{W_1^s(t),\ 0 \leq t \leq 1\}$. Then, by implicit function theorem, there is $k(s,W_1^s;\ \psi) \in K$ such that

$$F(s,k(s,W_1^s;\ \Phi_2(\sqrt{s}w^0)) + W_1^s,\ \Phi_2(\sqrt{s}w^0)) = v$$

with *sufficiently big probability*, (cf. Bismut [2], Kusuoka [7] and Malliavin [9]). Let

$$f(s,W_1^s;\ \psi) = \frac{1}{2}\|k(s,W^s;\ \psi)\|_H^2 + (\bar{h},k(s,W_1^s;\ \psi))_H \,.$$

Then there exists $g(s,W_1^s;\ \Phi_2(\sqrt{s}w^0))$ such that as $s \downarrow 0$

$$\tilde{q}(s,\underline{u},\underline{v}) \sim s^{-\frac{n}{2}} \exp[-\mathcal{E}[\gamma]/s] E^{\mu\otimes P_1^s}[g(s,W_1^s;\ \Phi_2(\sqrt{s}w^0))$$

$$\times \exp[-\frac{1}{s}f(s,W_1^s;\ \Phi_2(\sqrt{s}w^0)) - \frac{1}{\sqrt{s}}(\bar{h}^0,w^0)_{H^0} - \frac{1}{s}(\bar{h},W^s)_H]] \,.$$

$g(s,W_1^s;\ \Phi_2(\sqrt{s}w^0))$ is a good functional which serves only as a mollifier, (see Bismut [2], Chapter IV). Note that

$$f(0,h_1;\ \bar{\psi} + h^0) + (\bar{h},h_1)_H + (\bar{h}^0,h^0)_{H^0} + \frac{1}{2}\|h_1\|_H^2 + \frac{1}{2}\|h^0\|_{H^0}^2$$

$$= \mathcal{E}[\phi(\cdot;\ \bar{\psi} + h^0, h_1 + k(0,h_1;\ \bar{\psi} + h^0))]$$

and

$$X(1;\ 0,h_1 + k(0,h_1;\ \bar{\psi} + h^0) + \bar{h},\ \bar{\psi} + h^0) = v \,.$$

Here $\bar{\psi} = \Phi_2(0)$. Therefore we have

$$\partial_\psi f(0,0;\,\overline{\psi})(h^0) + Df(0,0;\,\overline{\psi})(h_1) + (\overline{h}^0, h^0)_{H^0} + (\overline{h}, h_1)_H$$

(4.6)

$$= \int_{t_*}^{t^*} h^0(t)N(t)\,dt$$

(cf. Lemma 2.1). Here $\partial_\psi f(0,0;\,\overline{\psi})$ is the usual Fréchet derivative of $f(0,0;\,\psi)$ at $\overline{\psi}$. By easy geometrical consideration, we see that $\ddot{\overline{\psi}}(t) \geqq 0$, $t \in [0,1]$. Then we have

$$\frac{1}{s}f(s,W^s;\,\Phi_2(\sqrt{s}w^0)) + \frac{1}{\sqrt{s}}(\overline{h}^0, w^0)_{H^0} + \frac{1}{s}(\overline{h}, W^s)_H$$

$$= \frac{1}{s}\int_0^1 (|\overline{\psi}(t) + \sqrt{s}w^0(t)| - \overline{\psi}(t) - \sqrt{s}w^0(t))\ddot{\overline{h}}^0(t)\,dt$$

(4.7) $$+ \frac{1}{\sqrt{s}}\int_{t_*}^{t^*} |w^0(t)|N(t)\,dt$$

$$+ \frac{1}{s}(\text{quadratic form of } W_1^s \text{ and } \Phi_2(\sqrt{s}w_0) - \overline{\psi})$$

$$+ \frac{1}{s}(\text{higher order}) .$$

Because the first term of the right hand side of (4.7) is positive and is dominated by

$$\frac{1}{\sqrt{s}}\int_{(t_*-\varepsilon,t_*)\cup(t^*,t^*+\varepsilon)} |w^0(t)|\ddot{\overline{h}}^0(t)\,dt$$

essentially, $\varepsilon > 0$, $N(t)$ is strictly positive for $t \in [t_*, t^*]$ and x and y are non-conjugate along γ,

$$\log \tilde{q}(s,\underline{u},\underline{v}) + \mathcal{E}[\gamma]/s$$

$$\sim \log E^\mu[\exp[-\frac{1}{\sqrt{s}}\int_{t_*}^{t^*} |w^0(t)|N(t)\,dt]] .$$

Since, as Lemma 4 in Ikeda [5], we obtain by the Feynman-Kac formula that

(4.8) $$-\log E^\mu[\exp[-\frac{1}{\sqrt{s}}\int_{t_*}^{t^*} |w^0(t)|N(t)\,dt]] \sim \lambda_1\int_{t_*}^{t^*} |N(t)|^{2/3}dt/(2s)^{1/3} .$$

(For detailed form of (4.8), see Hsu [4]). This implies the conclusion.

This research was partially supported by Grant-in-Aid for Scientific Research (No. 62302006), The Ministry of Education, Science and Culture.

References

[1] D. G. Aronson: Isolated singularities of solutions of second order parabolic equations, Arch. Ration. Mech. Anal., 19(1965), 231-238.

[2] J. M. Bismut: *Large deviations and the Malliavin calculus*, Progress in Math. 45, Birkhäuser, 1984.

[3] V. S. Buslaev: Continum integrals and the asymptotic behavior of the solution of parabolic equation as $t \to 0$. Application to diffraction, Topics in Math. Phys. 2(1968), ed. by M. Sh. Birman, 67-86.

[4] P. Hsu: Short time asymptotics of the heat kernel on concave boundary, preprint.

[5] N. Ikeda: On the asymptotic behavior of the fundamental solution of the heat equation on certain manifolds, Proc. Taniguchi Intern. Symp. on Stochastic Anal., Katata and Kyoto, 1982, ed. by K. Itô, Kinokuniya/North-Holland, 1984, 169-195.

[6] Y. Kannai: Off diagonal short time asymptotics for fundamental solutions of diffusion equations, Comm. Part. Diff. Equat., 2(1977), 781-830.

[7] S. Kusuoka: The generalized Malliavin calculus based on Brownian sheet and Bismut's expansion for large deviation, Lecture Notes in Math., 1158(1986), 141-157.

[8] P. Malliavin: Calcul des variations stochastiques subordonné au processus de la chaleur, C. R. Acad. Sc. Paris, Série I, 295(1982), 167-172.

[9] P. Malliavin: Implicit functions in finite corank on the Wiener space, Proc. Taniguchi Intern. on Stochastic Analy., Katata and Kyoto, 1982 ed. by K. Itô, Kinokuniya/North-Holland, 1984, 369-386.

[10] R. B. Melrose and M. E. Taylor: Near peak scattering and the corrected Kirchhoff approximation for a convex obstacle, Adv. in Math., 55(1985), 242-315.

[11] J. Milnor: *Morse theory*, Annals. Math. Studies, 51, Princeton Univ. Press, 1963.

[12] S. A. Molchanov: Diffusion processes and Riemannian geometry, Russian Math. Survey, 30(1975), 1-63.

[13] S. R. S. Varadhan: On the behavior of the fundamental solution of the heat equation with variable coefficients, Comm. Pure Appl. Math., 20(1967), 431-455.

MALLIAVIN CALCULUS ON A SEGAL SPACE

Kiyosi ITÔ
Research Institute for Mathematical Sciences
Kyoto University, Kyoto, 606 (Japan)

1. Introduction

Let $X = (X_\alpha, \alpha \in A)$ be a centered Gaussian system $\Omega = (\Omega, \mathcal{F}, P)$, for example, a finite or infinite dimensional Wiener process, a Brownian sheet or a white noise. In 1976 P.Malliavin [1] defined the C^∞ functions of X and opened up a new rich field, called Malliavin calculus, and S. Watanabe has recently defined the generalized functions of X. The customary way in this theory is to formulate definitions and theorems on the sample space of X, which is a Wiener space or more generally an abstract Wiener space. However, there are many ways of choosing the space. Instead of using the Wiener space representation we consider the closure (in $L_2(\Omega)$) of the given centered Gaussian system X, which turns out to be a Segal space (see Section 2 for the definition) and carry out discussions directly on the space $\Omega = (\Omega, \mathcal{F}, P)$.

Following S. Watanabe's Tata note [2] we define generalized functions by taking polynomials for test functions. We define differential forms directly and represent it as Hilbert space valued functions in the last section.

We published this idea in one of the IMA preprint series [3], where there were several clumsy points, which are improved here. This note is just a starting point. To go further we have to use Meyer's fundamental theorem. It can be carried out by reformulating arguments of Watanabe's note in our terminology.

We are very grateful to P. Malliavin, D. Stroock, S. Watanabe, I. Shigekawa and S. Kusuoka for their valuable discussions and comments.

2. Polynomials and generalized functions

A Segal space is defined to be a system of centered Gaussian variables on a probability space $\Omega = (\Omega, \mathcal{F}, P)$ closed under linear combinations and limits in probability.

Let S be a Segal space. The set of all $\sigma(S)$-measurable functions on Ω is denoted by $L_0(S)$. For $f \in L_0(S)$ we define

$$\|f\|_0 := E(|f| \wedge 1).$$

$\| \ \|_0$ is a quasinorm on $L_0(S)$ and the $\| \ \|_0$-convergence is equivalent to the convergence in probability. $(L_0(S), \| \ \|_0)$ is not only a Fréchet space but also a Fréchet ring. It is obvious that $S \cup \mathbb{R} \subset L_0(S)$.

Definition 2.1 The subring of $L_0(S)$ generated by $\mathbb{R} \cup S$ is denoted by $\mathcal{P}(S)$. An element of $\mathcal{P}(S)$ is called a polynomial on Ω relative to S or simply a polynomial on S.

For $p \in (0, \infty)$ we define

$$\|f\|_p := E(|f|^p)^{1/p} \in [0, \infty], \quad f \in L_0(S),$$

$$L_p(S) := \{f \in L_0(S) : \|f\|_p < \infty\}.$$

$(L_p(S), \|f\|_p)$ is a Banach space. As p increases, $\|f\|_p$ increases for every $f \in L_0(S)$ and so $L_p(S)$ decreases. Define

$$L_{\infty-}(S) := \bigcap_p L_p(S) = \lim_{p \to \infty} L_p(S),$$

$$\|f\|_{\infty-} := \sum_{n=1}^{\infty} 2^{-n}(\|f\|_n \wedge 1).$$

$(L_{\infty-}(S), \| \ \|_{\infty-})$ is not only a Fréchet space but also a Fréchet ring, as we can see from $\|fg\|_p \leqq \|f\|_{2p} \|g\|_{2p}$.

Since every $a \in S$ is centered Gaussian, it holds that $E(e^{c|a|}) < \infty$, $\forall c > 0$, We use this fact to prove the following:

Proposition 2.1 $\mathcal{P}(S)$ is a $\| \ \|_p$-dense subspace of $L_p(S)$ for $p \in \{0\} \cup [1, \infty) \cup \{-\infty\}$.

In this note p indicates a number in $(1, \infty)$ and q is the conjugate of p, i.e. $p^{-1} + q^{-1} = 1$, unless stated otherwise. If $f \in L_p(S)$, $g \in L_q(S)$, then $fg \in L_1(S)$ and $E(fg)$ is denoted by $\langle f,g \rangle$. It is obvious that

$$\|f\|_p = \sup\{|\langle f,g\rangle| \ : \ g \in L_q(S), \ \|g\|_q \leq 1\}.$$

Hence we obtain the following :

Proposition 2.2 If Q is a $\| \|_q$-dense linear subspace of $L_q(S)$ (for example, $Q = \mathcal{P}(S)$), then

$$\|f\|_p = \sup\{|\langle f,g\rangle| : g \in Q, \ \|g\|_q \leq 1\}.$$

Notation

N = the Gauss measure $N_{0,1}$.

$\mathbb{F}_{ON} = \mathbb{F}_{ON}(S)$ = the finite orthonormal systems in S.

$\mathcal{P}(\mathbb{R}^n)$ = the polynomials in $\mathbb{R}^n$.

$d(\varphi)$ = the degree of $\varphi \in \mathcal{P}(\mathbb{R}^n)$.

LS = linear span.

Let $\underline{e} = (e_1, e_2, \cdots, e_n) \in \mathbb{F}_{ON}$. Then we denote by $\mathcal{P}(\underline{e})$ the subring of $L_0(S)$ generated by $\mathbb{R} \cup \underline{e}$. It is obvious that $\mathcal{P}(\underline{e}) \subset \mathcal{P}(S)$. If $LS(\underline{e}) \subset LS(\underline{e}')$, $\underline{e}, \underline{e}'$ $\mathbb{F}_{ON}$, then $\mathcal{P}(\underline{e}) \subset \mathcal{P}(\underline{e})$. Also

$$\mathcal{P}(S) = \bigcup_{\underline{e} \in \mathbb{F}_{ON}} \mathcal{P}(\underline{e}).$$

Proposition 2.3 Let $\underline{e} = (e_1, e_2, \cdots, e_n) \in \mathbb{F}_{ON}$.

(i) $f \in \mathcal{P}(\underline{e})$

$\Leftrightarrow$ $f = \varphi(\underline{e})$, i.e. $f(\omega) = \varphi(\underline{e}(\omega))$ a.s., for some $\varphi \in \mathcal{P}(\mathbb{R}^n)$.

(ii) $\varphi(\underline{e}) = \psi(\underline{e})$, $\varphi, \psi \in \mathcal{P}(\mathbb{R}^n) \Rightarrow \varphi = \psi$.

Proof (i) is obvious. To prove (ii) note that $\underline{e}$ is N^n-distributed, so $\varphi = \psi$ a.e. (N^n) on $\mathbb{R}^n$ by the assumption, which implies that $\varphi = \psi$ everywhere on $\mathbb{R}^n$. ∎

Proposition 2.4 Let $\underline{a} = (a_1, a_2, \cdots, a_m)$, $\underline{b} = (b_1, b_2, \cdots, b_n) \in \mathbb{F}_{ON}$.

If $\varphi(\underline{a}) = \psi(\underline{b})$, $\varphi \in \mathcal{P}(\mathbb{R}^m)$, $\psi \in \mathcal{P}(\mathbb{R}^n)$, then

$$d(\varphi) = d(\psi).$$

<u>Proof</u> It suffices to prove that $d(\varphi) \leq d(\psi)$. Take an orthonormal base of $LS(a_1, a_2, \cdots, b_1, b_2, \cdots)$, $\underline{e} = (e_1, e_2, \cdots, e_\ell)$ such that $e_i = a_i$ for $i \leq m$. Since $b_j \in LS(\underline{e})$, we obtain

$$\varphi(\underline{a}) = \tilde{\varphi}(\underline{e}), \quad \psi(\underline{b}) = \tilde{\psi}(\underline{e}), \ d(\varphi) = d(\tilde{\varphi}), \ d(\psi) \geq d(\tilde{\psi}).$$

But $\tilde{\varphi}(\underline{e}) = \tilde{\psi}(\underline{e})$ by the assumption, so $\tilde{\varphi} = \tilde{\psi}$ by the last proposition (ii). Hence $d(\tilde{\varphi}) = d(\varphi)$, so $d(\varphi) \leq d(\psi)$. ∎

In view of this proposition we define the <u>degree</u> of $f \in \mathcal{P}(S)$, written $d(f)$ to be equal to $d(\varphi)$ if $f = \varphi(\underline{a})$, $\underline{a} = (a_1, a_2, \cdots, a_m) \in \mathbb{F}_{ON}$, $\varphi \in \mathcal{P}(\mathbb{R}^m)$.

Let

$$\mathcal{P}_k(S) := \{f \in \mathcal{P}(S) : d(f) \leq k\}.$$

$$\mathcal{P}_k(\underline{e}) := \{f \in \mathcal{P}(\underline{e}) : d(f) \leq k\}, \quad \underline{e} \in \mathbb{F}_{ON}$$

Then $\mathcal{P}_k(S) = \bigcup_{\underline{e}} \mathcal{P}_k(\underline{e})$.

Let $\mathcal{H}_k(\underline{e}) := \mathcal{P}_k(\underline{e}) \cap \mathcal{P}_{k-1}(\underline{e})$ $(k \geq 1)$, $\mathcal{H}_0(\underline{e}) = \mathcal{P}_0(\underline{e})$, where $\perp$ is the orthogonal complement in the Hilbert space $L_2(S)$. Being a finite dimensional subspace of $L_2(S)$, $\mathcal{P}_k(\underline{e})$ is a closed linear subspace of $L_2(S)$. Hence $\mathcal{P}_k(\underline{e}) = \mathcal{P}_{k-1}(e) \oplus \mathcal{H}_k(\underline{e})$(orthogonal sum), so

$$\mathcal{P}_k(\underline{e}) = \oplus_{i=1}^k \mathcal{H}_i(\underline{e}) \uparrow \mathcal{P}(\underline{e}) \tag{1}$$

<u>Proposition 2.5</u> Let $\underline{a}, \underline{b} \in \mathbb{F}_{ON}$. Then $\mathcal{H}_i(\underline{a}) \perp \mathcal{H}_j(\underline{b})$, $i \neq j$.

<u>Proof</u> We can assume that $i > j$. Let $f \in \mathcal{H}_i(\underline{a})$ and $g \in \mathcal{H}_j(\underline{b})$. Using the method used in the proof of proposition 2.4 , we can choose $\underline{e} = (\underline{a}, \underline{c}) \in \mathbb{F}_{ON}$ such that

$$f = \varphi(\underline{a}) \text{ and } g = \psi(\underline{e}) = \psi(\underline{a}, \underline{c})$$

where φ and ψ are polynomials. Since $g \in \mathcal{H}_j(\underline{b})$, $d(\psi) \leq j$. Hence we can represent $g = \psi(\underline{a}, \underline{c})$ in the form:

$$g = \Sigma_{k=1}^n \alpha_k(\underline{a}) \beta_k(\underline{c})$$

where α_k and β_k are polynomials of degree $\leq j$. Hence

$$fg = \sum_k \varphi(\underline{a})\alpha_k(\underline{a})\beta_k(\underline{c}).$$

Since $\underline{e} = (\underline{a}, \underline{c})$ is N^n-distributed, $\underline{a}$ and $\underline{c}$ are independent, so

$$E(fg) = \sum_k E(\varphi(\underline{a})\alpha_k(\underline{a}))E(\beta_k(\underline{c})).$$

Since $\varphi(\underline{a}) = f \in \mathcal{H}_i(\underline{a}), \alpha_k(\underline{a}) \in \mathcal{P}_j(\underline{a})$ and $j < i$, we obtain $E(\varphi(\underline{a})\ \alpha_k(\underline{a})) = 0$, so $E(fg) = 0$. ∎

Define

$$\mathcal{H}_k(S) = \bigcup_{\underline{e}} \mathcal{H}_k(\underline{e}) \tag{2}$$

Using (1) and the last proposition, we obtain the following:

<u>Theorem 2.1</u>

(i) $\mathcal{H}_i(S) \perp \mathcal{H}_j(S)$, $i \neq j$

(ii) $\mathcal{P}_k(S) = \oplus_{i=1}^k \mathcal{H}_i(S) \uparrow \mathcal{P}(S)$.

(iii) Every $f \in \mathcal{P}(S)$ is uniquely represented in the form:

$$f = \sum_{k=1}^{\infty} f_k, \quad f_k \in \mathcal{H}_k(S),$$

where $f_k = 0$ for $k > d(f)$.

<u>Definition 2.2</u>

$J_k f$ = the f_k of the theorem above.

$T_t f = \sum_{k=0}^{\infty} e^{-kt} J_k f, \quad t \in \mathbb{R}$.

$R_s f = \sum_{k=0}^{\infty} (1+k)^{-s/2} J_k f, \quad t \in \mathbb{R}$.

$Lf = -\sum_{k=0}^{\infty} k\ J_k f$.

Since $J_k f = 0$ for $k > d(f)$, the infinite sums above are essentially finite sums. $(J_k f(\omega), k=0,1,2,\cdots)$ is determined up to P-measure 0. Hence $T_t f(\omega)$ is continuous in t a.s. Similarly for $R_s f(\omega)$.

<u>Theorem 2.2</u>

(i) T_t, R_s and $I - L$ are linear bijections from $\mathcal{P}(S)$ to itself.

(ii) T_t, R_s, L : $\mathcal{H}_k(S) \to \mathcal{H}_k(S)$, $\mathcal{P}_k(S) \to \mathcal{P}_k(S)$.

(iii) $T_t T_s = T_t T_s$, $R_s R_t = R_{s+t}$, $T_t R_s = R_s R_t$, $T_t L = L T_t$, $R_s L = L R_s$.

(iv) $R_s f(\omega) = \int_0^\infty T_t f(\omega)\, \gamma_s(dt)$, $\forall s \geq 0$, a.s.

where $\gamma_s(dt) = \Gamma(s/2)^{-1} e^{-t} t^{s/2-1} dt (s > 0)$,

$= \delta_0(dt) (s = 0)$.

(v) $(T_t f, g) = (f, T_t g)$ and similarly for R_s and L.

Definition 2.3 For $\varphi \in \mathcal{P}(\mathbb{R}^n)$ we define

$$(L^{(n)}\varphi)(\underline{x}) := \sum_{i=1}^{n} (\partial_i^2 \varphi(\underline{x}) - x_i \partial_i \varphi(\underline{x})), \quad \underline{x} = (x_i) \in \mathbb{R}^n .$$

Using integration by parts, we obtain the following:

Proposition 2.6 $\langle L^{(n)}\varphi, \psi \rangle_n = \langle \varphi, L^{(n)}\psi \rangle_n$, where $\langle\, ,\rangle_n$ is the inner product in the Hilbert space $L_2(\mathbb{R}^n, N^n)$.

Proposition 2.7

(i) $d((L^{(n)} + k)\varphi) \leq d(\varphi)$.

(ii) $d(\varphi) \leq k \Rightarrow d((L^{(n)} + k)\ ) \leq k - 1$.

Proof (i) is obvious. Hence (ii) is obvious in case $d(\varphi) < k$. If $\varphi(\underline{x}) = x_1^{k_1} x_2^{k_2} \cdots x_n^{k_n}$ where $\sum_i k_i = k$, then $\sum x_i \partial_i \varphi = k\varphi$, so

$$d(L^{(n)}\varphi) = d(\textstyle\sum_i \partial_i^2 \varphi) < d(\varphi) < k.$$

Hence our proposition holds for every $\varphi \in \mathcal{P}(\mathbb{R}^n)$. ∎

Theorem 2.3 If $f = \varphi(\underline{e})$, $\underline{e} = (e_1, e_2, \cdots, e_n) \in \mathbb{F}_{ON}$, $\varphi \in \mathcal{P}(\mathbb{R}^n)$,

$$Lf = L^{(n)} \varphi(\underline{e}).$$

Proof It suffices to observe the case : $f \in \mathcal{H}_k(\underline{e})$. Then $Lf = -kf$ by the definition of L. Hence it suffices to prove that

$$\psi(\underline{e}) = 0 \quad \text{for} \quad \psi = (L^{(n)} + k)\varphi .$$

Since $f \in \mathcal{H}_k(\underline{e}) \in \mathcal{P}_k(\underline{e})$, we obtain $d(\varphi) = d(f) \leq k$. Hence $d(\psi) \leq k-1$ by the last proposition (ii). But

$$d((L^{(n)}+ k)\ ) \leq k-1 \quad \text{by the same proposition (i),}$$

i.e.

$$((L^{(n)}+ k)\psi)(\underline{e}) \in \mathcal{P}_{k-1}(\underline{e}), \text{ so } \langle f, (L^{(n)}+k)\psi(\underline{e})\rangle = 0.$$

Since $\underline{e}$ is N^n-distributed, we can use Proposition 2.6 to obtain

$$\begin{aligned}\langle \psi(\underline{e}), \psi(\underline{e})\rangle &= \langle \psi, \psi\rangle_n ,\\ &= \langle (L^{(n)}+ k)\varphi , \psi\rangle_n = \langle \varphi, (L^{(n)}+ k)\psi\rangle_n \\ &= \langle f, (L^{(n)}+ k)\psi(\underline{e})\rangle = 0,\end{aligned}$$

which implies that $\psi(\underline{e}) = 0$. ∎

Proposition 2.8 $\| T_t f \|_p \leq \| f \|_p$, $f \in \mathcal{P}$, $t \geq 0$.

Proof Representing f in the form $f = \varphi(\underline{e})$, $\underline{e} = \{e_1, \cdots, e_n\} \in \mathbb{F}_{ON}$ $\varphi \in \mathcal{P}(\mathbb{R}^n)$, we can prove that

$$T_t f = \int_{\mathbb{R}^n} p_t(\underline{e}, d\underline{y})\, \varphi(\underline{y}) ,$$

where $p_t(\underline{x}, \cdot)$ is the Gauss distribution on $\mathbb{R}^n$ with mean vector $e^{-t}\underline{x}$ and covariance matrix $(1-e^{-2t})(\delta_{ij})_{i,j=1}^{n}$. Noting that

$$N^n(d\underline{y}) = \int_{\underline{x} \in \mathbb{R}^n} N^n(d\underline{x})\, p_t(\underline{x}, d\underline{y}),$$

we obtain

$$E(|T_t f|^p) \leq \int_{\mathbb{R}^n} N^n(d\underline{y})|\varphi(y)|^p = E(|f|^p). \quad ∎$$

Proposition 2.9 $\| R_s f \|_p \leq \| f \|_p$, $f \in \mathcal{P}$, $s \geq 0$.

Proof Use Theorem 2.2 (iv) and the last proposition. ∎

Proposition 2.10

$$\| f \|_p = \sup\{ |\langle f,g\rangle| : g \in \mathcal{P}, \ \| g \|_q \leq 1\}, \quad f \in \mathcal{P}.$$

Proof Since $\mathcal{P} \subset L_p$, it is obvious that

$$\| f \|_p = \sup \{ |\langle f,g\rangle| : g \in L_q , \ \| g \|_q \leq 1\} .$$

But we can replace L_q by $\mathcal{P}$, because $\mathcal{P}$ is $\| \ \|_q$-dense in L_q

by $\mathcal{P}$, because $\mathcal{P}$ is $\| \|_q$-dense in L_q by Proposition 2.1. ∎

Now we define the <u>(s,p)-norm</u> $\|f\|_{s,p}$.

<u>Definition 2.4</u> $\|f\|_{s,p} = \|R_{-s}f\|_p$, $f \in \mathcal{P}$, $s \in \mathbb{R}$, $1 < p < \infty$.

It is obvious that $\|f\|_{0,p} = \|f\|_p$.

<u>Theorem 2.4</u> $\|f\|_{s,p}$ increases as s and p increase.

<u>Proof</u> Use $R_s R_t = R_{s+t}$ $(s,t \in \mathbb{R})$ and $\|R_s f\|_p \leqq \|f\|_p$ $(s \geqq 0)$. ∎

Let $\tilde{\mathcal{P}} = \tilde{\mathcal{P}}(S)$ be the linear space that consists of all linear maps : $\mathcal{P} = \mathcal{P}(S) \to \mathbb{R}$. For every $f \in \mathcal{P}$ we define $\pi f \in \tilde{\mathcal{P}}$ by

$$(\pi f)(g) = \langle f,g \rangle, \quad \forall g \in \mathcal{P} .$$

$\pi : \mathcal{P} \to \tilde{\mathcal{P}}$ is a linear injection. By identifying f with πf we regard $\mathcal{P}$ a linear subspace of $\tilde{\mathcal{P}}$, written $\mathcal{P} \subset \tilde{\mathcal{P}}$.

In view of the fact

$$(\pi R_s f)(g) = (R_s f,g) = (f,R_s g) = (\pi f)(R_s g)$$

we define $R_s F$ for $F \in \tilde{\mathcal{P}}$ as follows:

$$(R_s F)(g) := F(R_s g), \quad \forall g \in \mathcal{P} .$$

Similarly we define $T_t F$ and LF.

In view of Proposition 2.10 we define $\|F\|_p$ $(1 < p < \infty)$ for $F \in \tilde{\mathcal{P}}$ as follows:

$$\|F\|_p := \sup\{ |F(g)| : g \in \mathcal{P}(S), \ \|g\|_q \leqq 1\} \in [0,\infty] .$$

The (s,p)-norm of $F \in \tilde{\mathcal{P}}$ is defined as follows:

$$\|F\|_{s,p} := \|R_{-s}F\|_p , \quad s \in \mathbb{R} , \ p \in (1,\infty). \quad (\|F\|_{0,p} = \|F\|_p).$$

Since $\|g\|_q \downarrow$ as $q \downarrow$, $\|F\|_p \uparrow$ as $p \uparrow$. Since $R_t \mathcal{P} = \mathcal{P}$, we obtain $\|R_s F\|_p \leqq \|F\|_p$ for $s \geqq 0$ from Proposition 2.8. Noting that $R_s R_t = R_{s+t}$ we obtain the following:

<u>Theorem 2.5</u> $\|F\|_{s,p} \uparrow$ as $s,p \uparrow$.

<u>Definition 2.5</u> $\mathcal{D}_{s,p} = \mathcal{D}_{s,p}(S) := \{F \in \tilde{\mathcal{P}} : \|F\|_{s,p} < \infty\}$

$(\mathcal{D}_{s,p}, \| \|_{s,p})$ is a Banach space.

For $f \in L_p = L_p(S)$ we define $\pi f \in \tilde{\mathcal{P}}$ by

$$(\pi f)(g) = \langle f, g \rangle, \quad g \in \mathcal{P}.$$

This map $\pi : L_p(S) \to \tilde{\mathcal{P}}$ is a natural extension of the map $\pi : \mathcal{P} \to \tilde{\mathcal{P}}$ observed above. Since $\mathcal{P}$ is $\| \|_q$-dense in L_q, we obtain

$$\| \pi f\|_{0,p} = \| \pi f \|_p = \| f \|_p, \quad \pi(L_p) = \mathcal{D}_{0,p},$$

so $(\mathcal{D}_{0,p}, \| \|_{0,p})$ is isomorphic to $(L_p, \| \|_p)$ under $\pi : L_p \to \mathcal{D}_{0,p}$. Hence we identity $f \in L_p$ with $\pi f \in \mathcal{D}_{0,p}$, so $L_p = \mathcal{D}_{s,p}$.

<u>Theorem 2.6</u> $\mathcal{D}_{s,p} \downarrow$ as $s,p \uparrow$.

<u>Definition 2.6</u> $\mathcal{D} = \mathcal{D}(S) := \bigcap_{s,p} \mathcal{D}_{s,p} = \lim_{s,p\to\infty} \mathcal{D}_{s,p}$

An element of $\mathcal{D}\,(\subset \mathcal{D}_{0,p} = L_p)$ is called a <u>C_∞ function on Ω relative to S</u> (or simply <u>on S</u>) <u>in Malliavin's sense.</u>

<u>Definition 2.7</u> $\mathcal{D}' = \mathcal{D}'(S) := \bigcup_{s,p} \mathcal{D}_{s,p} = \lim_{\substack{s\to-\infty \\ p\to 1}} \mathcal{D}_{s,p}$.

An element of $\mathcal{D}'$ is called a <u>generalized function on Ω relative to S</u> (or simply <u>on S</u>) <u>in S. Watanabe's sense</u>.

If $s > 0$ and $1 < p < \infty$, then

$$\mathcal{P} \subset \mathcal{D}_{s,p} \subset \mathcal{D}_{0,p} = L_p \subset \mathcal{D}_{-s,p} \subset \mathcal{D}' \subset \tilde{\mathcal{P}}.$$

If $s > 0$, then $\mathcal{D}_{s,p}(\subset L_{0,p})$ consists of functions and $\mathcal{D}_{-s,p}$ contains proper generalized functions.

<u>3. Polynomial vector fields</u>

We denote $L_p(S)$ and $\mathcal{P}(S)$ simply by L_p and $\mathcal{P}$.

<u>Definition 3.1</u> A linear map $V : \mathcal{P} \to \mathcal{P}$ is called a <u>polynomial vector field on Ω relative to S</u> (or simply <u>on S</u>), written

$$V \in \mathcal{J} = \mathcal{J}(S),$$

if the following conditions are satisfied:

(V.1) (derivation property) $V(fg) = V(f)g + fV(g)$

(V.2) (property of finite rank) There exists a finite dimensional subspace of S, $F = F_V$ such that

$$V(e) = 0 \quad \text{whenever} \quad e \perp F.$$

If $V_1, V_2 \in \mathcal{J}$, then $f_1V_1 + f_2V_2 \in \mathcal{J}$, where

$$(f_1V_1 + f_2V_2)(g) := f_1V_1(g) + f_2V_2(g), \ \forall g \in \mathcal{P}.$$

Theorem 3.1 $\mathcal{J}$ is a $\mathcal{P}$-module.

Theorem 3.2 If $\varphi \in \mathcal{P}(\mathbb{R}^n)$ and $\underline{f} = (f_1, f_2, \cdots, f_n)$, $f_i \in \mathcal{P}$, then

$$V(\varphi(\underline{f})) = \Sigma_{i=1}^n (\partial_i\varphi)(\underline{f})V(f_i), \quad V \in \mathcal{J}.$$

Since every $f \in \mathcal{P}$ is represented in the form:

$$f = \varphi(\underline{a}), \ \underline{a} = (a_1, a_2, \cdots, a_n) \in \mathbb{F}_{ON}, \ \varphi \in \mathcal{P}(\mathbb{R}^n)$$

it holds that

$$V(f) = \Sigma_{i=1}^n (\partial_i\varphi)(\underline{a})V(a_i), \quad V \in \mathcal{J}.$$

Hence we obtain the following:

Theorem 3.3 $V_1 = V_2$ on $S \Rightarrow V_1 = V_2 \ (V_1 V_2 \in \mathcal{J})$.

Definition 3.2 For $e \in S$ we define the vector field in the direction e, written D_e, by the condition :

$$D_e \in \mathcal{J}, \ D_e(a) = \langle a, e \rangle, \ \forall a \in S.$$

The existence and the uniqueness of such D_e are easily verified.

Theorem 3.4 Every $V \in \mathcal{J}$ is represented in the form:

$$V = \Sigma_{i=1}^n f_i D_{e_i},$$

where $\{e_i\}$ is an orthonormal base (abbr. ONB) of F_V((V.2) in

Definition 3.1) and $f_i = V(e_i) \in \mathcal{P}$.

Theorem 3.5

(i) $e \mapsto D_e$ is linear from S into $\mathcal{J}$.

(ii) $D_a D_b = D_b D_a$. (Note that $V_1 V_2 \neq V_2 V_1$ in general).

4. Polynomial r-forms

Let $\mathcal{J}^r$ be the product space of r copies of $\mathcal{J}$ for $r=1,2,3,\cdots$.

Definition 4.1 A map $\alpha : \mathcal{J}^r \to \mathcal{P}$ is called a polynomial r-form on Ω relative to S (or simply on S), if the following conditions are satisfied:

(α.1) ($\mathcal{P}$-multilinearity) For each $i = 1,2,\cdots,r$ the map $V_i \mapsto \alpha(V_1,V_2,\cdots, V_r)$ is $\mathcal{P}$-linear when all $V_j (j \neq i)$ are fixed.

(α.2) (Property of finite rank) There exists a finite dimensional subspace of S, $F = F_\alpha$ such that

$$\alpha(D_{h_1}, D_{h_2}, \cdots, D_{h_r}) = 0,$$

whenever $h_i \perp F$ for at least one index i.

We denote the polynomial r-forms by $p^{\otimes r} = p^{\otimes r}(S)$ or simply by $\mathcal{P}^r = \mathcal{P}^r(S)$. We define

$$\mathcal{P}^0 := \mathcal{P}$$

for convenience.

Theorem 4.1 $\mathcal{P}^r$ is a $\mathcal{P}$-module.

Theorem 4.2 Let $\alpha, \beta \in \mathcal{P}^r$. If

$$\alpha(D_{h_1}, D_{h_2}, \cdots, D_{h_r}) = \beta(D_{h_1}, D_{h_2}, \cdots, D_{h_r}), \quad \forall h_i \in S,$$

then $\alpha = \beta$.

Proof Use (α.1) and Theorem 3.4. ∎

Definition 4.2 For $f \in \mathcal{P}^0$ we define the gradient of f, $\nabla f \in \mathcal{P}^1$ as follows:

$$(\nabla f)(V) = V(f), \quad \forall V \in \mathcal{J}.$$

Theorem 4.3 $\nabla : \mathcal{P}^0 \to \mathcal{P}^1$ is linear and satisfies the derivation property:

$$\nabla(fg) = \nabla f \cdot g + f \cdot \nabla g$$

Definition 4.3 For $\alpha \in \mathcal{P}^r$ and $\beta \in \mathcal{P}^s$ we define their tensor product $\alpha \otimes \beta \in \mathcal{P}^{r+s}$ as follows:

$$(\alpha \otimes \beta)(V_1, V_2, \cdots, V_{r+s}) := \alpha(V_1, \cdots, V_r)\beta(V_{r+1}, \cdots, V_{r+s}).$$

Theorem 4.3

(i) $\alpha \otimes \beta$ is $\mathcal{P}$-bilinear in (α, β).

(ii) $(\alpha \otimes \beta) \otimes \gamma = \alpha \otimes (\beta \otimes \gamma)$. Hence $\alpha \otimes \beta \otimes \gamma$ and $\alpha_1 \otimes \alpha_2 \otimes \cdots \otimes \alpha_r$ are meaningful. Note that $\alpha \otimes \beta \neq \beta \otimes \alpha$ in general.

Theorem 4.5 Every $\alpha \in \mathcal{P}^2$ is represented in the form:

$$\alpha = \Sigma_{i,j=1}^{n} f_{ij} \nabla e_i \otimes \nabla e_j,$$

where $\{e_1, e_2, \cdots, e_n\}$ is an ONB of F_α(see(α.2)) and $f_{i,j} = \alpha(D_{e_i}, D_{e_j}) \in \mathcal{P}$. Similarly for $\alpha \in \mathcal{P}^r (r \neq 2)$.

The definition given below are independent of the representation, though we do not mention it.

Definition 4.4 Suppose that $\alpha, \beta \in \mathcal{P}^2$ are represented as

$$\alpha = \Sigma f_{i_1 i_2} \nabla a_{i_1} \otimes \nabla a_{i_2}, \quad \beta = \Sigma g_{j_1 j_2} \nabla b_{j_1} \otimes \nabla b_{j_2},$$

where i_1 and i_2 run over $\{1, 2, \cdots, m\}$, and j_1 and j_2 run over $\{1, 2, \cdots, n\}$. Then we define $\alpha \cdot \beta \in \mathcal{P}$ by

$$\alpha \cdot \beta := \Sigma f_{i_1 i_2} g_{j_1 j_2} \langle a_{i_1}, b_{j_1} \rangle \langle a_{i_2}, b_{j_2} \rangle .$$

Similarly we define $\alpha \cdot \beta \in \mathcal{P}$ for $\alpha, \beta \in \mathcal{P}^r$ $(r \neq 2)$.

Definition 4.5

$$\langle \alpha, \beta \rangle := E(\alpha \cdot \beta).$$

Theorem 4.6

(i) $\alpha\cdot\beta$ is $\mathcal{P}$-linear and symmetric in (α,β), and $\alpha\cdot\alpha \geq 0$ where the equality holds only for $\alpha=0$.

(ii) $\langle\alpha,\beta\rangle$ is linear and symmetric in (α,β) and $\langle\alpha,\alpha\rangle \geq 0$ where the equality holds only for $\alpha=0$.

Definition 4.6 The divergence of $\alpha \in \mathcal{P}^r$, $\nabla^*\alpha$ is defined as follows: If α is represented in the form

$$\alpha = \sum f_{i_1 i_2 \cdots i_r} \nabla a_{i_1} \otimes \nabla a_{i_2} \otimes \cdots \otimes \nabla a_{i_r},$$

$$(\{a_1, a_2, \cdots, a_m\} \in \mathbb{F}_{ON})$$

then

$$\nabla^*\alpha := \sum (\nabla f_{i_1 \cdots i_r} \cdot \nabla a_{i_1} - f_{i_1 \cdots i_r} a_{i_1}) \nabla a_{i_2} \otimes \cdots \otimes \nabla a_{i_r}.$$

Theorem 4.7 $\langle \nabla^*\alpha, \beta \rangle = \langle \alpha, \nabla\beta \rangle$, $\alpha \in \mathcal{P}^r$, $\beta \in \mathcal{P}^{r-1}$.

Proof Represent α in the form above and β in the form:

$$\beta = \sum g_{j_1 j_2 \cdots j_r} \nabla b_{j_1} \otimes \nabla b_{j_2} \otimes \cdots \otimes \nabla b_{j_r}$$

$$(\{b_1, b_2, \cdots, b_n\} \in \mathbb{F}_{ON})$$

Let $\underline{e} = \{e_1, \cdots, e_m, e_{m+1}, \cdots, e_\ell\}$ be an ONB of the linear span of $a_1, \cdots, a_m, b_1, \cdots, b_n$ and all elements of S appearing in the representations of $f\ldots$ and $g\ldots$. We can assume that $e_i = a_i$, $i \leq m$. Representing α and β in terms of $\underline{e}$, we can express $\langle \nabla^*\alpha, \beta \rangle$ and $\langle \alpha, \nabla\beta \rangle$ as integrals on $(\mathbb{R}^\ell, N^\ell)$, because $\underline{e}$ is N^ℓ-distributed. We can use integration by parts to check these two integrals are equal. ∎

If $f = \varphi(\underline{e})$, $\underline{e} = \{e_1, \cdots, e_n\} \in \mathbb{F}_{ON}$, $\varphi \in \mathcal{P}(\mathbb{R}^n)$, then

$$\nabla^*\nabla f = \nabla^*(\textstyle\sum_i \partial_i\varphi(\underline{e}) \nabla e_i)$$

$$= -\sum_i (\nabla \partial_i \varphi(\underline{e}) \cdot \nabla e_i - e_i \partial_i \varphi(\underline{e}))$$

$$= -\sum_i (\partial_i^2 \varphi(\underline{e}) - e_i \partial_i \varphi(\underline{e})) \quad (\nabla e_i \cdot \nabla e_j = (e_i, e_j) = \delta_{ij})$$

$$= -(L_n \varphi)(\underline{e}).$$

This, combined with Theorem 2.3, proves the following:

Theorem 4.8 $Lf = -\nabla^* \nabla f$ for $f \in \mathcal{P}$.

In view of this theorem we define $L\alpha$ for $\alpha \in \mathcal{P}^r$ as follows:

Definition 4.7 $L\alpha := -\nabla^* \nabla \alpha$ for $\alpha \in \mathcal{P}^r$.

Theorem 4.9 If α is represented in the form of Definition 4.6, then

$$L\alpha = \sum L f_{i_1 \cdots i_r} \nabla e_{i_1} \otimes \cdots \otimes \nabla e_{i_r} .$$

Definition 4.8 If α is represented in the form of Definition 4.6, then we define

$$T_t \alpha := \sum T_t f_{i_1 \cdots i_r} \nabla e_{i_1} \otimes \cdots \otimes \nabla e_{i_r} , \quad t \in \mathbb{R} ,$$

and similarly for $R_s \alpha$, $s \in \mathbb{R}$.

Theorem 4.10 (i), (iii), (iv) and (v) of Theorem 2.2 hold on $\mathcal{P}^r = \mathcal{P}^r(S)$.

Definition 4.9 For $\alpha \in \mathcal{P}^r$ we define $|\alpha| \in L_{\infty -}$ is follows : $|\alpha| := (\alpha \cdot \alpha)^{1/2}$.

Definition 4.10 $\|\alpha\|_p = \| |\alpha| \|_p$, $\alpha \in \mathcal{P}^r$, $\forall p \in [1, \infty)$

Theorem 4.11 Suppose that $\alpha, \beta \in \mathcal{P}^r$.

(i) $|\alpha| \geqq 0$ (the equality holds only for $\alpha = 0$),

$|\alpha + \beta| \leqq |\alpha| + |\beta|$, $|f\alpha| = |f| |\alpha|$ $(f \in \mathcal{P})$,

$|\alpha \cdot \beta| \leqq |\alpha| |\beta|$.

(ii) $\|\alpha\|_p \geqq 0$ (the equality holds only for $\alpha = 0$),

$\|\alpha + \beta\|_p \leqq \|\alpha\|_p + \|\beta\|_p$,

$|\langle \alpha, \beta \rangle| \leqq \|\alpha\|_p \|\beta\|_q$.

(iii) $\|\alpha\|_p = \sup\{|\langle \alpha, \beta \rangle| : \beta \in \mathcal{P}^r, \|\beta\|_q \leqq 1\}$.

Proof (iii) is the only statement that needs proof.

The inequality $\leq$ is obvious by $|\langle\alpha,\beta\rangle| \leq \|\alpha\|_p \|\beta\|_q$. Let us prove the opposite inequality. To avoid notational complications we observe the case : r=1. Then α is represented in the form:

$$\alpha = \sum_{i=1}^{n} f_i \nabla e_i, \quad \underline{e} = (e_1, e_2, \cdots, e_n) \in \mathbb{F}_{ON}, \quad f_i \in \mathcal{P}.$$

Since our assertion holds trivially in case $f_i = 0$, $\forall i$, we assume that $f = (\sum f_i^2)^{1/2} > 0$. Let

$$g_i := f_i \, f^{(p-q)/q} \, \|f\|_p^{-p/q}$$

$$(\sum g_i^2)^{1/2} = f^{p/q} / \|f\|_p^{p/q}, \quad \text{so} \quad \|(\sum g_i^2)^{1/2}\|_q = 1.$$

$$\sum f_i g_i = f^p / \|f\|_p^{p/q}, \quad \text{so} \quad E(\sum f_i g_i) = \|f\|_p$$

Since $g_i \in L_q$, we can find a sequence $\{g_{i,n}\} \subset \mathcal{P}$ such that

$$\|(\sum g_{i,n}^2)^{1/2}\|_q = 1 \quad \text{and} \quad E(\sum f_{i,n}\, g_{i,n}) > \|f\|_p - n^{-1}.$$

Let $\beta_n : \sum g_{i,n} \nabla e_n$. Then $\|\beta_n\|_q = 1$ and

$$\langle\alpha,\beta\rangle = E(\alpha \cdot \beta_n) > \|f\|_p - n^{-1}.$$

Hence the right hand side of (iii) is $\geq \|f\|_p - n^{-1}$ for every n and so $\to \|f\|_p$. ∎

Definition 4.11 $\|\alpha\|_{s,p} = \|R_{-s}\alpha\|_p$. $s \in \mathbb{R}$, $1 < p < \infty$ is called the (s,p)-norm of $\alpha \in \mathcal{P}^r$. $\|\alpha\|_{0,p} = \|\alpha\|_p$ obviously.

Theorem 4.12 $\|\alpha\|_{s,p}$ increases as s and p increase.

5. Generalized r-forms

Let $\tilde{\mathcal{P}}^{r*} = \tilde{\mathcal{P}}^{r\otimes}(S)$ be the set of all linear maps from $\mathcal{P}^r$ into $\mathbb{R}$, where $r = 0,1,2,\cdots$. For simplicity we denote $\mathcal{P}^{r\otimes}$ simply by $\mathcal{P}^r$ in this note. Let $A \in \tilde{\mathcal{P}}^r$. The value of A at $\beta \in \mathcal{P}^r$ is denoted by $A[\beta]$. It is obvious that $\tilde{\mathcal{P}}^r$ is a linear space.

For every $\alpha \in \mathcal{P}^r$ we define $\pi\alpha \in \tilde{\mathcal{P}}^r$ as follows:

$$\pi\alpha(\beta) := \langle \alpha,\beta \rangle , \quad \forall\beta \in \mathcal{P}^r.$$

The map $\pi : \mathcal{P}^r \to \tilde{\mathcal{P}}^r$ is a linear injection. By identifying $\pi\alpha$ with α we regard $\mathcal{P}^r$ a linear subspace of $\tilde{\mathcal{P}}^r$, so

$$\mathcal{P}^r \subset \tilde{\mathcal{P}}^r .$$

Since

$$\pi(R_s\alpha)(\beta) = \langle R_s\alpha,\beta \rangle = \langle \alpha,R_s\beta \rangle = \pi\alpha(R_s\beta),$$

we define $R_sA \in \tilde{\mathcal{P}}^r$ for $A \in \tilde{\mathcal{P}}^r$ as follows:

$$R_sA(\beta) := A(R_s\beta), \quad \forall\beta \in \mathcal{P}^r , \quad s \in \mathbb{R}.$$

Similarly we define T_tA and LA for $A \in \tilde{\mathcal{P}}^r$. Then $R_s, T_t I$ I-L : $\tilde{\mathcal{P}}^r \to \tilde{\mathcal{P}}^r$ are linear bijections. In view of Theorem 4.11 (iii) and Theorem 4.7 we define $\|A\|_p$ $(A \in \tilde{\mathcal{P}}^r)$, $\nabla: \tilde{\mathcal{P}}^r \to \tilde{\mathcal{P}}^{r+1}$ and $\nabla^*: \tilde{\mathcal{P}}^r \to \tilde{\mathcal{P}}^{r+1}$ as follows:

$$\|A\|_p := \sup\{ |\langle \alpha,\beta \rangle| : \beta \in \mathcal{P}^r, \|\beta\|_q \leq 1\},$$

$$\nabla A(\beta) := A(\nabla^*\beta), \quad \forall\beta \in \mathcal{P}^{r+1} ,$$

$$\nabla^*A(\beta) := A(\nabla\beta), \quad \forall\beta \in \mathcal{P}^r .$$

<u>Theorem 5.1</u> $\|A\|_p$ increases as p increases.

<u>Definition 5.1</u> $\|A\|_{s,p} = \|R_{-s}A\|_p$, $A \in \tilde{\mathcal{P}}^r$.

$$\mathcal{D}^{r\otimes}_{s,p} = \mathcal{D}^{r\otimes}_{s,p}(S) := \{A \in \tilde{\mathcal{P}}^r : \|A\|_{s,p} < \infty\}.$$

$$\mathcal{D}^{r\otimes} = \mathcal{D}^{r\otimes}(S) := \bigcap_{s,p} \mathcal{D}^{r\otimes}_{s,p}$$

$$\mathcal{D}'^{r\otimes} = \mathcal{D}'^{r\otimes}(S) := \bigcup_{s,p} \mathcal{D}^{r\otimes}_{s,p}$$

Hereafter we omit $\otimes$ in the symbols above.

<u>Theorem 5.2</u> As r and p increase, $\|A\|_{s,p}$ increases, so

$$\mathcal{D}^r_{s,p} \downarrow \mathcal{D}^r \quad (r \uparrow \infty, \ p \uparrow \infty)$$

$$\mathcal{D}^r_{s,p} \uparrow \mathcal{D}'^r \quad (r \downarrow -\infty, \ p \downarrow 1).$$

Definition 5.2 An element of $\mathcal{D}^r$ is called a C^∞ r-form on Ω relative to S (or simply on S) in Malliavin's sense.

Definition 5.3 An element of $\mathcal{D}'$ is called a generalized r-form on Ω relative to S (or simply on Ω) in S. Watanabe's sense.

6. Regular representations of r-forms

Let H be a Hilbert space isomorphic to S under the map $\theta : H \to S$. An H-valued function Φ on Ω is called $\sigma(S)$-measurable, written

$$\Phi \in L_0(\Omega \to H, \sigma(S), P) \quad \text{or simply} \quad \Phi \in L_0(S,H),$$

if $\langle \Phi(\omega), h\rangle_H$ is $\sigma(S)$-measurable in $\omega \in \Omega$. An H-valued function Φ on Ω is called an H-valued polynomial, written

$$\Phi \in \mathcal{P}(S, H) ,$$

if Φ is represented in the form :

$$\Phi(\omega) = \Sigma_{i=1}^n f_i(\omega)\, h_i \quad (f_i \in \mathcal{P}(S), \ h_i \in H) \quad \text{a.s.}$$

It is obvious that

$$\mathcal{P}(S,H) \subset L_0(S,H).$$

Let $\{h_i, i \in I\}$ be an ONB in H. Then for every $\Phi \in L_0(S,H)$ we obtain

$$\Phi(\omega) = \Sigma_{i\in I} f_i(\omega) h_i, \quad f_i(\omega) = \langle \Phi(\omega), h_i\rangle ,$$

where $f_i(\omega) \equiv 0$ except for a countable number of i's, so $\|\Phi\|_H(\omega) := \|\Phi(\omega)\|_H = (\Sigma_i f_i(\omega)^2)^{1/2}$ is $\sigma(S)$-measurable in ω, namely $\|\Phi\|_H \in L_0(S)$.

Definition 6.1

$$\|\Phi\|_p := \| \|\Phi\|_H \|_p \ (\in [0,\infty]), \quad \Phi \in L_0(S,H),$$

$$L_p(S,H) := \{\Phi \in L_0(S,H) : \|\Phi\|_p < \infty\}.$$

$(L_p(S,H), \| \|_p)$ is a Banach space.

Proposition 6.1 $\mathcal{P}(S,H)$ is a $\| \|_p$-dense subspace of $L_p(S,H)$.

Proof Let $\Phi \in L_p(S,H)$. Then for $\varepsilon > 0$ we can find $\Phi_\varepsilon(\omega) = \sum_{i=1}^{n(\varepsilon)} f_{i,}(\omega)\, h_{i,\varepsilon}$ $(f_{i,\varepsilon} \in L_p(S),\ h_{i,\varepsilon} \in H)$ such that $\|\Phi_\varepsilon - \Phi\|_p < \varepsilon$. But $\mathcal{P}(S)$ is $\| \|_p$-dense in $L_p(S)$ by Proposition 2.1. Hence Φ is $\| \|_p$-approximated by elements of $\mathcal{P}(S,H)$. ▮

For $\Phi \in L_p(S,H)$ and $\Psi \in L_q(S,H)$, we define

$$\langle \Phi, \Psi \rangle = E(\langle \Phi, \Psi \rangle_H),$$

Noting that

$$E(|\langle \Phi, \Psi\rangle_H|) = E(\|\Phi\|_H \|\Psi\|_H) \leqq \|\Phi\|_p \|\Psi\|_q .$$

It is obvious that

$$|\langle \Phi, \Psi \rangle| \leqq \|\Phi\|_p \|\Psi\|_q .$$

By the technique used in the proof of Theorem 4.11 (iii) we obtain the following:

Proposition 6.2 Let $\Phi \in L_p(S,H)$. Then

$$\|\Phi\|_p = \sup\{|\langle \Phi, \Psi\rangle| : \Psi \in \mathcal{P}(S,H),\ \|\Psi\|_p \leqq 1\}.$$

Proposition 6.3 If $\Phi \in L_p(S,H)$, then $\ell_\Phi(\Psi) := \langle \Phi, \Psi \rangle$ is linear and $\| \|_q$-continuous in $\Psi \in \mathcal{P}(S,H)$. Conversely every linear and $\| \|_q$-continuous map $\ell : \mathcal{P}(S,H) \to \mathbb{R}$ is represented as $\ell(\Psi) = \langle \Phi, \Psi \rangle$ $(\Phi \in L_p(S,H))$.

Let $\Psi \in \mathcal{P}(S,H)$. Then Ψ is represented as

$$\Psi(\omega) = \sum_{i=1}^n f_i(\omega)\, h_i \quad \text{a.s.}, \quad f_i \in \mathcal{P}(S), \quad h_i \in H.$$

We define $\Theta\Psi \in \mathcal{P}^1(S)$ as follows :

$$\Theta\Psi := \sum_{i=1}^{n} f_i \nabla e_i, \quad e_i = \theta h_i .$$

This is well-defined independently of the representation of Ψ.

Proposition 6.4 $(\mathcal{P}^1(S), \| \|_p)$ is isomorphic to $(\mathcal{P}(S,H), \| \|_p)$ under $\Theta : \mathcal{P}(S,H) \to \mathcal{P}^1(S)$. If $\Phi, \Psi \in \mathcal{P}(S,H)$, then

$$(\Theta\Phi \cdot \Theta\Psi)(\omega) = \langle \Phi, \Psi \rangle_H(\omega) \quad \text{a.s.,}$$

and so

$$\langle \Theta\Phi, \Theta\Psi \rangle = \langle \Phi, \Psi \rangle ,$$

i.e.

$$\langle \Theta\Phi, \beta \rangle = \langle \Phi, \Theta^{-1}\beta \rangle , \quad \beta \in \mathcal{P}^1(S).$$

In view of this fact we extend $\Theta : \mathcal{P}(S,H) \to \mathcal{P}^1(S) (\subset \mathcal{D}^1_{0,p}(S))$ to a map from $L_p(S,H) \to \tilde{\mathcal{P}}^1(S)$ as follows :

$$(\Theta\Phi)[\beta] = \Phi[\Theta^{-1}\beta] \quad \forall \beta \in \mathcal{P}^1(S).$$

Using Propositions 6.4, 6.3 and 6.2, we obtain the following :

Theorem 6.1 If $\Phi \in \mathcal{P}(S,H)$, then

$$\|\Theta\Phi\|_p = \|\Phi\|_p , \quad \text{so } \Theta\Phi \in \mathcal{D}^1_{0,p}(S).$$

Hence $(\mathcal{D}^1_{0,p}(S), \| \|_p)$ is isomorphic to $(L_p(S,H)), \| \|_p)$ under $\Theta : L_p(S,H) \to \mathcal{D}^1_{0,p}(S)$.

Hence every $A \in \mathcal{D}^1_{0,p}(S)$ is represented in the form:

$$A = \Theta\Phi, \quad \Phi \in \mathcal{P}(S,H).$$

Definition 6.2 This representation is called a regular representation of A.

Since $\Theta(\mathcal{P}(S,H)) = \mathcal{P}^1(S) \subset \mathcal{D}^1_{0,p}(S)$, this theorem, combined with Proposition 6.1, implies the following :

Theorem 6.2 $\mathcal{P}^1(S)$ is $\| \|_{0,p}$-dense in $\mathcal{D}^1_{0,p}(S)$.

Definition 6.3

(i) For $F \in L_p(S) (\subset \mathcal{D}_{0,p}(S))$ and $e \in S$ we define $F \nabla e \in \tilde{\mathcal{P}}^1(S)$ as follows:

$$(F \nabla e)[\beta] := \langle F, \nabla e \cdot \beta \rangle, \quad \forall \beta \in \mathcal{P}^1(S),$$

(ii) For $A \in \mathcal{D}^1_{0,p}(S)$ and $e \in S$ we define $A \cdot \nabla e \in \tilde{\mathcal{P}}(S)$ as follows:

$$(A \cdot \nabla e)[f] = A[f \nabla e], \quad \forall f \in \mathcal{P}(S).$$

Theorem 6.3

(i) $\|F \cdot \nabla e\|_p = \|F\|_p \|e\|_2$, so $F \cdot \nabla e \in \mathcal{D}^1_{0,p}(S)$.

(ii) $\|A \cdot \nabla e\|_p \leq \|A\|_p \|e\|_2$, so $A \cdot \nabla e \in L_p(S)$.

Proof $|\langle F, \nabla e \cdot \beta \rangle| = \|F\|_p \|\nabla e \cdot \beta\|_q$. But $|\nabla e \cdot \beta| \leq |\nabla e| |\beta| \leq \|e\|_2 |\beta|$. This proves that $\|F \cdot \nabla e\|_p \leq \|F\|_p \|e\|_2$. If $\beta = f \nabla e$, $f \in \mathcal{P}(S)$, then $\nabla e \cdot \beta = f$ and $\|\beta\|_q = \|f\|_q$. Hence $\|F \cdot \nabla e\|_p = \|F\|_p \|e\|_2$. This complete the proof of (i). (ii) follows from $\|f \nabla e\|_q = \|f\|_q$. ∎

Theorem 6.4 Suppose that $e = \theta h$. Then

(i) $\Theta(Fh) = F \nabla e$, $F \in L_p(S)$.

(ii) $\langle \Phi, h \rangle_H(\omega) = (\Theta \Phi \cdot \nabla e)(\omega)$ a.s., $\Phi \in L_p(S,H)$.

Proof (i) If $\beta = \sum f_i \nabla e_i$, $e_i = \theta h_i$, then $\Theta^{-1}\beta = \sum f_i h_i$. Then

$$\Theta(F \cdot h)[\beta] = \Sigma_i \langle F, f_i \rangle (h, h_i) = (F \nabla e)[\beta].$$

This proves (i). To prove (ii) observe that

$$[\Theta \Phi \cdot \nabla e](f) = (\Theta \Phi)[f \nabla e] = \langle \Phi, fh \rangle$$

$$= E(\langle \Phi, fh \rangle_H) = E(f \langle \Phi, h \rangle_H)$$

$$= \langle \langle \Phi, h \rangle_H, f \rangle, \quad \forall f \in \mathcal{P}(S).$$

Hence $\Theta \Phi \cdot \nabla e = \langle \Phi, h \rangle_H$. ∎

<u>Theorem 6.5</u> Let $\{e_i, i \in I\}$ be an ONB in S. Then every $A \in \mathcal{D}'_{0,p}$ in represented in the form :

$$A = \sum_{i\in I} F_i \cdot \nabla e_i \quad (\text{in } \| \ \|_{0,p}\text{-convergence})$$

where

$$F_i = A \cdot \nabla e_i$$

<u>Proof</u> Take $H = \ell_2(I)$ and

$$\theta(h) = \sum_i a_i e_i \ , \qquad h = (a_i).$$

Then $\theta(h_i) = e_i$, $h_i = (\delta_{ij})_{j\in I}$.
Let

$$\Phi := \Theta^{-1} A \in L_p(S,H)$$

Then

$$\Phi(\omega) = \sum_i F_i(\omega)\, h_i \quad (\text{in } \| \ \|_H\text{-convergence})$$

and

$$\Phi = \sum_i F_i\, h_i \quad (\text{in } \| \ \|_p\text{-convergence})$$

where $F_i = \langle \Phi, h_i \rangle_H$, i.e. $\Phi = (F_i)_{i\in I}$.
But

$$A = \Theta \Phi = \sum_i \Theta(F_i h_i) = \sum_i F_i \nabla e_i$$

and

$$F_i(\omega) = (\Theta \Phi \cdot \nabla e)(\omega) = (A \cdot \nabla e)(\omega),$$

as we can check by using the facts proved above. ∎

<u>Theorem 6.6</u> Let $(\Lambda, \mathcal{B}, m)$ be a measure space, and $\tilde{\mathcal{B}} := \{M \in \mathcal{B} \ m(M) < \infty\}$. Suppose that $\{e(M), M \in \tilde{\mathcal{B}}\}$ satisfies the following conditions:

(W.1) $e(M) \in S$,

(W.2) $CLS\{e(M) \in \tilde{\mathcal{B}}\} = S$,

(W.3) $\langle e(M), e(M')\rangle = m(M \cap M')$.

Then every $A \in \mathcal{D}^1_{0,p}(S)$ is represented as

$$A = \int_\Lambda f_\lambda(\omega)\, \nabla e(d\lambda).$$

<u>Proof</u> Let $H = L_2(\Lambda, \mathcal{B}, m)$ and

$$\theta(\varphi) = \int_\Lambda \varphi(\lambda)\, e(d\lambda) \quad \text{(Wiener integral)}$$

for $\varphi \in H$. Let $1_M(\lambda)$ be the indicator of the set $M \in \tilde{\mathcal{B}}$. Let $\Phi = \Theta^{-1}A$.

Then

$$\Phi = (\Phi_\lambda(\omega),\ \lambda \in \Lambda,\ \omega \in \Lambda) \in L_p(S, \Lambda),$$

and

$$(\Phi_\lambda(\omega), \lambda \in \Lambda) \in H, \quad \forall \omega.$$

Hence

$$\Phi_\cdot(\omega) = \int_\Lambda \Phi_\lambda(\omega)\, 1_{d\lambda}(\cdot)$$

Applying Θ to both sides of this equation, we obtain the representation of A above. ∎

Similar observation can be made for $\mathcal{D}^r_{0,p}$ by using $H^{r\otimes}$ instead of H and for $\mathcal{D}^r_{s,p}$ by using $R_s : \mathcal{D}^r_{0,p} \to \mathcal{D}^r_{s,p}$.

References

(1) P.Malliavin : Stochastic calculus of variation and hypoelliptic operators, Proc. Int. Symp. on Stochastic Diferential equations, Kyoto (1976), Kinokuniya, 195-263.

(2) S.Watanabe : Lectures on stochastic differential equations, Tata Institute Lecture Note 73 (1984).

[3] K.Itô : Malliavin's C^∞ functionals of a centered Gaussian system. IMA preprint Series, Univ. of Minnesota, No. 327 (1987).

WEAK CONVERGENCE OF FUNCTIONALS OF POINT PROCESSES ON R^d

Yuji KASAHARA
Institute of Mathematics. University of Tsukuba.
Sakuramura. Ibaraki, 305 (Japan)

Makoto MAEJIMA
Department of Mathematics. Faculty of Science and Technology
Keio University, Hiyoshi, Yokohama, 223 (Japan)

§1. INTRODUCTION

In recent works ([7,8]) the authors used the point process method to prove some limit theorems for processes which may be expressed as weighted sums of i.i.d. random variables. The present paper is their continuation and we shall treat the case of multi-parameter.

A real-valued stochastic process $X=\{X(\underset{\sim}{t});\underset{\sim}{t}\in R_+^d\}$ $(d\geq 1)$, with multi-dimensional time parameter, is said to be self-similar with parameter $H(\in R)$ if for every $c>0$, $\{X(c\underset{\sim}{t})\}_{\underset{\sim}{t}}$ and $\{c^H X(\underset{\sim}{t})\}_{\underset{\sim}{t}}$ have the same finite-dimensional distributions.

The fractional stable process (or the fractional stable sheet when $d\geq 2$) (see Section 4 for the definition) has been considered as one of important classes of self-similar processes. One reason why self-similar processes are important and well-studied is that all limiting processes of time-space scaled processes are self-similar. Hence, once one finds a self-similar process, the next natural question is to find its domain of attraction. In the case $d=1$, there are some literatures on limit theorems of the fractional stable process (Astrauskas [1], Avram-Taqqu [2], Kasahara-Maejima [8], Maejima [10]). Here we present a limit theorem for the fractional stable sheet with $d\geq 2$. For that purpose, we shall again use the point process method on R^d.

In Section 2, we define the stochastic integral with respect to the stable sheet (the Lévy stable process with multi-dimensional time parameter). In Section 3, we introduce point processes induced by a set of independent, identically distributed multi-indexed random variables and show weak convergence of their certain functionals. In Section 4, we apply the result in Section 3 to find the domain of attraction of the fractional stable sheet.

§2. PRELIMINARIES

Let $\xi(dx)$ be a stable law on R with index α $(0<\alpha<2)$. In the case

$\alpha=1$ we further assume that $\xi(dx)$ is symmetric. Let $\nu(dx)$ be the Lévy measure of $\xi(dx)$. Thus $\nu(dx)$ is a Borel measure on $R\setminus\{0\}$ which is of the form

$$\nu(dx) = \alpha\{C_+ I[x>0] + C_- I[x<0]\} \, |x|^{-\alpha-1} \, dx,$$

where C_+, $C_- \geq 0$, $C_+ + C_- > 0$ and, if $\alpha=1$, $C_+ = C_-$.

Let $N(d\underset{\sim}{u}\, dx)$ be a Poisson random measure on $R^d \times (R\setminus\{0\})$ $(d \geq 1)$ with mean measure $\hat{N}(d\underset{\sim}{u}\, dx) = \mu(d\underset{\sim}{u})\nu(dx)$, where $\mu(d\underset{\sim}{u})$ denotes the Lebesgue measure on R^d. Following Ikeda-Watanabe [5], let us denote

$$\tilde{N}(d\underset{\sim}{u}\, dx) = N(d\underset{\sim}{u}\, dx) - \hat{N}(d\underset{\sim}{u}\, dx).$$

For a Borel measurable function $f: R^d \to R$ and $c>0$, we define

$$M_c(f) = \iint_{|f(\underset{\sim}{u})x| \leq c} f(\underset{\sim}{u})x \, \tilde{N}(d\underset{\sim}{u}\, dx),$$

$$A_c(f) = \iint_{|f(\underset{\sim}{u})x| > c} f(\underset{\sim}{u})x \, N(d\underset{\sim}{u}\, dx)$$

and

$$a_c(f) = \begin{cases} \displaystyle\iint_{|f(\underset{\sim}{u})x| \leq c} f(\underset{\sim}{u})x \, \hat{N}(d\underset{\sim}{u}\, dx), & \text{if } 0<\alpha<1, \\ 0 \;, & \text{if } \alpha=1, \\ \displaystyle -\iint_{|f(\underset{\sim}{u})x| > c} f(\underset{\sim}{u})x \, \hat{N}(d\underset{\sim}{u}\, dx), & \text{if } 1<\alpha<2. \end{cases}$$

Here, $M_c(f)$ is defined in the sense of stochastic integral and, therefore, is well-defined if

$$\iint_{|f(\underset{\sim}{u})x| \leq c} (f(\underset{\sim}{u})x)^2 \, \hat{N}(d\underset{\sim}{u}\, dx) < \infty.$$

$A_c(f)$ and $a_c(f)$ are defined in the sense of usual Lebesgue integrals provided that they exist (a.s.).

We finally put

$$X(f) = M_c(f) + A_c(f) + a_c(f)$$

when all of the three terms on the right-hand side are well-defined. The reader should remark that formally $X(f)$ may be expressed as

$$X(f) = \begin{cases} \displaystyle\iint_{R^d \times (R\setminus\{0\})} f(\underset{\sim}{u})x \, N(d\underset{\sim}{u}\, dx), & \text{if } 0<\alpha<1, \\ \displaystyle\iint_{R^d \times (R\setminus\{0\})} f(\underset{\sim}{u})x \, \tilde{N}(d\underset{\sim}{u}\, dx), & \text{if } 1\leq\alpha<2. \end{cases}$$

Therefore, it is easy to see that $X(f)$ does not depend on $c>0$ and, hence, we shall choose c so that

(2.1) $$\hat{N}(\{(\underset{\sim}{u},x);|f(\underset{\sim}{u})x|=c\}) = 0,$$

unless stated otherwise explicitly.

THEOREM 2.1. If

$$\int_{R^d} |f(\underset{\sim}{u})|^\alpha \mu(d\underset{\sim}{u}) < \infty,$$

then, all of $M_c(f)$, $A_c(f)$, $a_c(f)$ and hence $X(f)$ are well-defined.

Proof. Since

$$\int\int_{|f(\underset{\sim}{u})x|\le c} (f(\underset{\sim}{u})x)^2 \hat{N}(d\underset{\sim}{u}\, dx)$$

$$= (C_+ + C_-)\, c^{2-\alpha}\alpha/(2-\alpha) \int_{R^d} |f(\underset{\sim}{u})|^\alpha \mu(d\underset{\sim}{u}) < \infty,$$

$M_c(f)$ is well-defined. Next, notice that

$$\int\int_{|f(\underset{\sim}{u})x|>c} \hat{N}(d\underset{\sim}{u}\, dx) = (C_+ + C_-)c^{-\alpha} \int_{R^d} |f(\underset{\sim}{u})|^\alpha \mu(d\underset{\sim}{u}) < \infty.$$

This proves that $I[|f(\underset{\sim}{u})x|>c]\cdot N(d\underset{\sim}{u}\, dx)$ is a finite sum of Dirac measures (a.s.), which proves that $A_c(f)$ is well-defined. In a similar way we can easily see that so is $a_c(f)$.

Remark. The above definition $X(f)$ of stochastic integral is essentially the same as that given by Taqqu-Wolpert [11]. The only difference is that, when $\alpha=1$, they do not assume $C_+=C_-$ and instead they put an additional condition

$$\int_{R^d} \left|f(\underset{\sim}{u}) \log|f(\underset{\sim}{u})|\right| \mu(d\underset{\sim}{u}) < \infty.$$

Example 1. For $\underset{\sim}{t}=(t_1,\cdots,t_d)\in R^d_+$, let

$$f_{\underset{\sim}{t}}(\underset{\sim}{u}) = I[0\le u_1\le t_1, \cdots, 0\le u_d\le t_d], \qquad \underset{\sim}{u}=(u_1,\cdots,u_d)\in R^d.$$

Then, $Z_{(\alpha)}(\underset{\sim}{t})=X(f_{\underset{\sim}{t}})$ is the so called α-stable random field (or α-stable sheet).

Example 2. (Taqqu-Wolpert [11]) Let $\gamma_1, \cdots, \gamma_d$ be real numbers such that

(2.2) $$-\frac{1}{\alpha} < \gamma_j < 1-\frac{1}{\alpha}, \quad \gamma_j \ne 0 \quad (j=1,\cdots,d),$$

and let

$$f_{\tau}(u) = \prod_{j=1}^{d} \left[((u_j-\tau_j)^-)^{\gamma_j} - (u_j^-)^{\gamma_j} \right], \qquad \tau=(\tau_1,\cdots,\tau_d)\in R_+^d,\ u\in R^d.$$

where $x^- = \max\{-x,0\}$. (We define $0^{\gamma}=0$ even if $\gamma<0$.) Then $\{X(f_{\tau});\tau\in R_+^d\}$ is well-defined and is called the fractional stable sheet (or fractional Lévy motion with multi-parameter). A generalization will be given in Section 4.

THEOREM 2.2. *There exists a constant* C (>0) *depending only on* $\nu(dx)$ *such that*

$$P[\ |X(f)|>a\ ] \le C(a^{-2}+1) \int_{R^d} |f(u)|^{\alpha}\, \mu(du)$$

for every $a>0$ *and every Borel measurable function* $f(u)$ *on* R^d *such that the right-hand side is finite.*

Proof. Since the proof is essentially the same as that of Theorem 3.2 in the next section, we omit the proof.

§3. POINT PROCESSES INDUCED BY I.I.D. RANDOM VARIABLES AND THEIR CONVERGENCE

Let $\xi(dx)$ be the same as before and let $F(x)$ be the distribution function of a law belonging to the domain of attraction of $\xi(dx)$. In other words there exists a positive $\psi(x)$ ($\uparrow \infty$ as $x \to \infty$) such that

$$n\, dF(\psi(n)x) \longrightarrow \nu(dx), \quad \text{as} \quad n \longrightarrow \infty, \tag{3.1}$$

vaguely on $[-\infty, \infty]\backslash\{0\}$, (measures on $(-\infty, \infty)\backslash\{0\}$ are always extended to those of $[-\infty, \infty]\backslash\{0\}$ in the usual way) supplemented by the condition

$$\begin{cases} \int x\, dF(x) = 0, & \text{when } 1<\alpha<2, \\ \lim_{n\to\infty} n\int_{|x|<1} x\, dF(\psi(n)x) = 0, & \text{when } \alpha=1. \end{cases} \tag{3.2}$$

Let $\{\xi_i;\ i\in Z^d\}$ be independent random variables with distribution function $F(x)$ above, and define a sequence of integer-valued random measures $\{N_n(du\, dx)\}_n$ on $R^d\times(R\backslash\{0\})$ as follows:

$$N_n(du\, dx) = \sum_{i;\xi_i\neq 0} \delta_{(i/n,\,\xi_i/\psi(n^d))}(du\, dx),$$

where $\delta_{(u,x)}$ denotes the Dirac measure located at (u,x).

THEOREM 3.1. *Let* $N(d\underset{\sim}{u}\,dx)$ *be as in Section 2. Then,*

$$N_n(d\underset{\sim}{u}\,dx) \xrightarrow{\mathcal{L}} N(d\underset{\sim}{u}\,dx), \quad \text{as} \quad n \to \infty,$$

with respect to the vague topology of $R^d\times([-\infty,\infty]\setminus\{0\})$.

(Throughout the paper $\xrightarrow{\mathcal{L}}$ means the convergence in law.)

Proof. Let

$$\mu_n(d\underset{\sim}{u}) = n^{-d}\sum_{\underset{\sim}{k}\in Z^d}\delta_{\underset{\sim}{k}/n}(d\underset{\sim}{u}), \qquad \text{on } R^d. \tag{3.3}$$

Then it holds that $\mu_n \to \mu$, as $n \to \infty$, vaguely on R^d, μ being the Lebesgue measure. Therefore, denoting

$$\nu_n(dx) = n^d\, dF(\psi(n^d)x), \tag{3.4}$$

we have from (3.1) that $E[N_n(d\underset{\sim}{u}\,dx)]$ $(= \mu_n(d\underset{\sim}{u})\nu_n(dx))$ converges to $E[N(d\underset{\sim}{u}\,dx)]$ $(= \mu(d\underset{\sim}{u})\nu(dx))$ as $n \to \infty$, vaguely on $R^d\times([-\infty,\infty]\setminus\{0\})$. Now it is easy to conclude the assertion by Poisson's law of small numbers. (See also the proof of Theorem 3.1 of Durrett-Resnick [4].)

By the definition of the vague topology, the assertion of Theorem 3.1 is equivalent to

$$\int\!\!\int g(\underset{\sim}{u},x)\, N_n(d\underset{\sim}{u}\,dx) \xrightarrow{\mathcal{L}} \int\!\!\int g(\underset{\sim}{u},x)\, N(d\underset{\sim}{u}\,dx)$$

for every continuous function $g(\underset{\sim}{u},x)$ with compact support in $R^d\times([-\infty,\infty]\setminus\{0\})$, (see Theorem 3 of Jagers [6]). Also note that if $\{g_n(\underset{\sim}{u},x)\}_n$ and $g(\underset{\sim}{u},x)$ are measurable functions on $R^d\times((-\infty,\infty)\setminus\{0\})$ vanishing outside a common compact subset of $R^d\times([-\infty,\infty]\setminus\{0\})$, and if $g_n(\underset{\sim}{u},x)$ converges continuously to $g(\underset{\sim}{u},x)$, $\mu(d\underset{\sim}{u})\cdot\nu(dx)$-a.e., then

$$\int\!\!\int g_n(\underset{\sim}{u},x)\, N_n(d\underset{\sim}{u}\,dx) \xrightarrow{\mathcal{L}} \int\!\!\int g(\underset{\sim}{u},x)\, N(d\underset{\sim}{u}\,dx), \ (n \to \infty). \tag{3.5}$$

(This can easily be shown by using the argument in the proof of Lemma 6.3 of Kasahara-Watanabe [9].)

We now define

$$\begin{aligned} X_n(f) &= \int\!\!\int f(\underset{\sim}{u})x\, N_n(d\underset{\sim}{u}\,dx) \\ &= \frac{1}{\psi(n^d)}\sum_{\underset{\sim}{i}\in Z^d} f(\underset{\sim}{i}/n)\xi_{\underset{\sim}{i}}, \qquad n\geq 1, \end{aligned} \tag{3.6}$$

for any measurable function $f(\underset{\sim}{u})$ on R^d such that the right-hand side of (3.6) converges in probability. In the rest of this section, we shall discuss the convergence of $\{X_n(f)\}_n$ as n tends to infinity. The difficulty is that this problem cannot directly be reduced to (3.5) because integrands $g_n(\underset{\sim}{u},x) = f_n(\underset{\sim}{u})x$ and $g(\underset{\sim}{u},x) = f(\underset{\sim}{u})x$ do not vanish in the neighbourhood of x-axis, in general, and hence do not have compact support in $R^d \times ([-\infty,\infty] \setminus \{0\})$. Thus we need a cut-off method and to this end we prepare the following basic inequality.

THEOREM 3.2. *Let $\varepsilon > 0$ and let $f(\underset{\sim}{u})$ be a measurable function on R^d such that*

$$\sup_n \int_{R^d} \{|f(\underset{\sim}{u})|^{\alpha+\varepsilon} + |f(\underset{\sim}{u})|^{\alpha-\varepsilon}\} \mu_n(d\underset{\sim}{u}) < \infty,$$

where μ_n is defined in (3.3). Then, $X_n(f)$ $(n \geq 1)$ are well-defined and we have

$$(3.7) \qquad P[\,|X_n(f)| > a\,] \leq C(a^{-2}+1) \int_{R^d} \{|f(\underset{\sim}{u})|^{\alpha+\varepsilon} + |f(\underset{\sim}{u})|^{\alpha-\varepsilon}\} \mu_n(d\underset{\sim}{u})$$

for every $a > 0$, where $C(>0)$ is a constant which is independent of a, n, and f.

For the proof we prepare

LEMMA 3.1. *Let μ_n and ν_n be as in (3.3)-(3.4), and for given $\varepsilon(>0)$ and $\delta(\in(0,\infty])$, we put*

$$C_{\varepsilon,\delta,n} = \int_{|x|\leq\delta} \min\{|x|^{\alpha+\varepsilon}, |x|^{\alpha-\varepsilon}\} \nu_n(dx).$$

(i) *If $0 \leq \beta < \alpha$, then, for every $n \geq 1$,*

$$\iint_{|f(\underset{\sim}{u})x|>1,\,|x|\leq\delta} |f(\underset{\sim}{u})x|^\beta \mu_n(d\underset{\sim}{u})\nu_n(dx)$$

$$\leq C_{\varepsilon,\delta,n} \int_{R^d} \{|f(\underset{\sim}{u})|^{\alpha+\varepsilon} + |f(\underset{\sim}{u})|^{\alpha-\varepsilon}\} \mu_n(d\underset{\sim}{u}).$$

(ii) *If $\alpha < \beta$, then, for every $n \geq 1$,*

$$\iint_{|f(\underset{\sim}{u})x|\leq 1,\,|x|\leq\delta} |f(\underset{\sim}{u})x|^\beta \mu_n(d\underset{\sim}{u})\nu_n(dx)$$

$$\leq C_{\varepsilon,\delta,n} \int_{R^d} \{|f(\underset{\sim}{u})|^{\alpha+\varepsilon} + |f(\underset{\sim}{u})|^{\alpha-\varepsilon}\} \mu_n(d\underset{\sim}{u}).$$

Proof. Since this lemma is an easy modification of Lemma 2.1 of [8], the details are left to the reader.

Proof of Theorem 3.2. Let $c \in (0,1]$ and we define

$$M_c^n(f) = \int\int_{|f(\underset{\sim}{u})x| \le c} f(\underset{\sim}{u})x \; (N_n - E[N_n])(d\underset{\sim}{u}\, dx),$$

$$A_c^n(f) = \int\int_{|f(\underset{\sim}{u})x| > c} f(\underset{\sim}{u})x \; N_n(d\underset{\sim}{u}\, dx)$$

and

$$a_c^n(f) = \int\int_{|f(\underset{\sim}{u})x| \le c} f(\underset{\sim}{u})x \; E[N_n(d\underset{\sim}{u}\, dx)].$$

Thus we have

$$X_n(f) = M_c^n(f) + A_c^n(f) + a_c^n(f).$$

Now we shall show that the inequality (3.7) holds with replacing $X_n(f)$ by each of $M_c^n(f)$, $A_c^n(f)$ and $a_c^n(f)$. We first consider $A_c^n(f_n)$. Applying Lemma 3.1 with $\beta=0$, we have

(3.8)
$$\begin{aligned} & P[\iint_{|f(\underset{\sim}{u})x| > c} N_n(d\underset{\sim}{u}\, dx) \ge 1] \\ & \le E[\iint_{|f(\underset{\sim}{u})x| > c} N_n(d\underset{\sim}{u}\, dx)] \\ & = \iint_{|f(\underset{\sim}{u})x| > c} \mu_n(d\underset{\sim}{u}) \nu_n(dx) \\ & \le C_{\varepsilon,\infty,n} \int_{R^d} \{ |f(\underset{\sim}{u})|^{\alpha+\varepsilon} + |f(\underset{\sim}{u})|^{\alpha-\varepsilon} \} \mu_n(d\underset{\sim}{u}). \end{aligned}$$

This proves that $I[|f(\underset{\sim}{u})x| > c] \cdot N_n(d\underset{\sim}{u}\, dx)$ is in fact a finite sum of Dirac measures (a.s.), and hence, $A_c^n(f_n)$ is well-defined. Since

$$\lim_{n \to \infty} C_{\varepsilon,\infty,n} = \int \min\{ |x|^{\alpha+\varepsilon}, |x|^{\alpha-\varepsilon} \} \nu(dx) \; (< \infty),$$

(see [8]) we have that $\{C_{\varepsilon,\infty,n}\}_n$ remains bounded as $n \to \infty$. Keeping in mind that

$$P[\iint_{|f(\underset{\sim}{u})x| > c, |x| < \delta} f(\underset{\sim}{u})x \; N_n(d\underset{\sim}{u}\, dx) \ne 0]$$

is less than or equal to the extreme left-hand side of (3.8), we also see that (3.7) holds, $X_n(f)$ being replaced by $A_c^n(f)$. Similarly, by the martingale inequality and Lemma 3.1 (ii) with $\beta=2$, we have

(3.9)
$$P[|M_c^n(f)| > a] \le a^{-2} E[(M_c^n(f))^2]$$

$$\leq a^{-2} \iint_{|f(\underset{\sim}{u})x| \leq c} (f(\underset{\sim}{u})x)^2 \, \mu_n(d\underset{\sim}{u}) \nu_n(dx)$$

$$\leq a^{-2} C_{\varepsilon,\infty,n} \int_{R^d} \{|f(\underset{\sim}{u})|^{\alpha+\varepsilon} + |f(\underset{\sim}{u})|^{\alpha-\varepsilon}\} \mu_n(d\underset{\sim}{u}).$$

Since $a_c^n(f)$ may be treated in the same way, we omit the details.

THEOREM 3.3. *Let $f_n(\underset{\sim}{u})$ $(n \geq 1)$ and $f(\underset{\sim}{u})$ be real-valued measurable functions on R^d satisfying the following three conditions:*

(A1) *$f_n(\underset{\sim}{u})$ converges continuously to $f(\underset{\sim}{u})$, $\mu(d\underset{\sim}{u})$-a.e.,*

(A2) *for every $T>0$, there exists a $\beta(>\alpha)$ such that*

$$\sup_n \int_{|\underset{\sim}{u}| \leq T} |f_n(\underset{\sim}{u})|^{\beta} \mu_n(d\underset{\sim}{u}) < \infty,$$

and

(A3) *there exists an $\varepsilon_0(>0)$ such that*

$$\lim_{T \to \infty} \limsup_{n \to \infty} \int_{|\underset{\sim}{u}|>T} \{|f_n(\underset{\sim}{u})|^{\alpha-\varepsilon_0} + |f_n(\underset{\sim}{u})|^{\alpha+\varepsilon_0}\} \mu_n(d\underset{\sim}{u}) = 0.$$

Then $X_n(f_n)$ and $X(f)$ are well-defined, and also

$$X_n(f_n) \xrightarrow{\mathcal{L}} X(f), \quad \text{as } n \to \infty.$$

Proof. Let $M_c^n(\cdot)$, $A_c^n(\cdot)$ and $a_c^n(\cdot)$ be as in the proof of Theorem 3.2, and we shall show that $M_c^n(f_n)$, $A_c^n(f_n)$ and $a_c^n(f_n)$ converge jointly to $M_c(f)$, $A_c(f)$ and $a_c(f)$, for every c $(0<c\leq 1)$ satisfying (2.1). For that, it is enough to show the joint convergence of N_n with each of $M_c^n(f_n)$, $A_c^n(f_n)$ and $a_c^n(f_n)$, respectively. We first show

$$(\, M_c^n(f_n),\, N_n \,) \xrightarrow{\mathcal{L}} (\, M_c(f),\, N \,), \quad \text{as } n \to \infty.$$

For the moment let us assume that f_n vanishes outside a (common) bounded subset of R^d, i.e., $f_n(\underset{\sim}{u}) = 0$ if $|\underset{\sim}{u}|>T$. Let $\delta \in (0,\infty)$. By (3.5) and (A1) we have

$$(3.10) \quad \left(\iint_{|f_n(\underset{\sim}{u})x| \leq c, |x|>\delta} f_n(\underset{\sim}{u})x \, (N_n - E[N_n])(d\underset{\sim}{u}\, dx),\ N_n(d\underset{\sim}{u}\, dx) \right)$$

$$\xrightarrow{\mathcal{L}} \left(\iint_{|f(\underset{\sim}{u})x| \leq c, |x|>\delta} f(\underset{\sim}{u})x \, \tilde{N}(d\underset{\sim}{u}\, dx),\ N(d\underset{\sim}{u}\, dx) \right),$$

for every $c(\in (0,1])$ satisfying (2.1). Here we have used that $g_n(\underset{\sim}{u},x) = I[|f_n(\underset{\sim}{u})x| \leq c, |x|>\delta] \cdot f_n(\underset{\sim}{u})x$ is uniformly bounded and converges to $g(\underset{\sim}{u},x)=I[|f(\underset{\sim}{u})x|\le c,|x|>\delta]\cdot f(\underset{\sim}{u})x$, a.e. ((2.1) is crucial but the condition $c\le 1$ is not.) On the other hand, with a slight modification of (3.9), we obtain that

$$(3.11)\quad P(\delta,n) \equiv P[\ |\int\int_{|f_n(\underset{\sim}{u})x|\le c,|x|\le\delta} f_n(\underset{\sim}{u})x\ (N_n-E[N_n])(d\underset{\sim}{u}\ dx)| > a]$$

$$\le a^{-2}\ C_{\varepsilon,\delta,n}\int_{|\underset{\sim}{u}|\le T}\{|f_n(\underset{\sim}{u})|^{\alpha+\varepsilon}+|f_n(\underset{\sim}{u})|^{\alpha-\varepsilon}\}\mu_n(d\underset{\sim}{u}).$$

If we choose $\varepsilon(>0)$ so that $\alpha+\varepsilon<\beta$, we have from (A2) and Hölder's inequality that

$$P(\delta,n) \le \text{const. } C_{\varepsilon,\delta,n}$$

and therefore

$$(3.12)\qquad \lim_{\delta\downarrow 0}\ \limsup_{n\to\infty}\ P(\delta,n)$$

$$\le \text{const.} \lim_{\delta\downarrow 0}\int_{|x|\le\delta}\min\{|x|^{\alpha+\varepsilon},|x|^{\alpha-\varepsilon}\}\ \nu(dx) = 0.$$

Now it is a standard argument (see Theorem 4.2 of [3]) that (3.10) combined with (3.12) (also Theorem 2.2) implies

$$(\ M_c^n(f_n),\ N_n) \xrightarrow{\ \mathcal{L}\ } (\ M_c(f),\ N\ ),\ (n\to\infty).$$

In a similar way, using Theorem 3.2 and (A3), we can remove the restriction that $f_n(\underset{\sim}{u})$ vanishes outside $\{\underset{\sim}{u};|\underset{\sim}{u}|\le T\}$.

The convergence in law of $(A_c^n(f_n),\ N_n)$ and $a_c^n(f)$ can be proved in the same line. In the proof for $A_c^n(f_n)$, (3.11) will be replaced by an obvious modification of (3.8), and condition (3.2) will be used for the convergence of $a_c^n(f_n)$. Since all necessary ideas are found in [8], we omit the details.

§4. THE DOMAIN OF ATTRACTION OF FRACTIONAL STABLE SHEETS

For real a, b, τ and γ $(\gamma\neq 0)$, we shall denote

$$f_{a,b,\gamma}(\tau;u) = a\{((u-\tau)^-)^\gamma - (u^-)^\gamma\} + b\{((u-\tau)^+)^\gamma - (u^+)^\gamma\},\quad u\in R,$$

where $x^+=\max\{x,0\}$, $x^-=(-x)^+$. (We define $0^\gamma=0$ even if $\gamma<0$.) By Theorem 2.1, we easily have

LEMMA 4.1. *Let a_i, b_i and γ_i $(i=1,\cdots,d)$ be real numbers. If (2.2) is satisfied, then for every $\underset{\sim}{\tau}=(\tau_1,\cdots,\tau_d)\in R_+^d$,*

(4.1) $$\Delta(\underset{\sim}{\tau}) = X\left(\prod_{i=1}^{d} f_{a_i,b_i,\gamma_i}(\tau_i;u_i)\right)$$

is well-defined.

We call the random field $\{\Delta(\underset{\sim}{\tau});\ \underset{\sim}{\tau}\in R_+^d\}$ defined by (4.1) the fractional stable sheet or the fractional Lévy motion with multi-parameter. A special case is already given in [11], (see Example 2 in Section 2). It is not difficult to see that $\{\Delta(\underset{\sim}{\tau})\}$ is a self-similar random field with parameter $H = (d/\alpha)+\gamma_1+\cdots+\gamma_d$. This fact also follows from Theorem 4.1 below. In the present section we shall apply Theorem 3.3 to find its domain of attraction, i.e., stationary multi-indexed sequences whose partial sums converge in law to $\{\Delta(\underset{\sim}{\tau})\}_{\underset{\sim}{\tau}}$ under a suitable normalization.

Let a_i, b_i and γ_i $(i=1,\cdots,d)$ be as in Lemma 4.1 and let $\{c_j^{(i)}\}_{j=-\infty}^{\infty}$ $(i=1,\cdots,d)$ be sequences of real numbers such that, for every i, there exists a $\psi_i(x)$ $(x\geq 0)$ which varies regularly at ∞ with exponent γ_i. We further assume;

(B1) $$c_n^{(i)} = O(\psi_i(|n|)/|n|) \quad \text{as} \quad |n| \longrightarrow \infty,$$

(B2-1) for $i(1\leq i\leq d)$ such that $\gamma_i>0$, it holds

(i) $$\lim_{n\to\infty} \frac{1}{\psi_i(n)} \sum_{0\leq j\leq n} c_j^{(i)} = a_i\ ,$$

(ii) $$\lim_{n\to\infty} \frac{1}{\psi_i(n)} \sum_{-n\leq j<0} c_j^{(i)} = -b_i,$$

(B2-2) for $i(1\leq i\leq d)$ such that $\gamma_i<0$, it holds

(i) $$\lim_{n\to\infty} \frac{1}{\psi_i(n)} \sum_{j>n} c_j^{(i)} = -a_i,$$

(ii) $$\lim_{n\to\infty} \frac{1}{\psi_i(n)} \sum_{j<-n} c_j^{(i)} = b_i$$

and

(iii) $$\sum_j c_j^{(i)} = 0.$$

Now let $\{\xi_{\underset{\sim}{i}};\ \underset{\sim}{i}\in Z^d\}$ be as in Section 3, and define a stationary sequence $\{Y_{\underset{\sim}{k}};\ k_1,\cdots,k_d\geq 1\}$ of random variables by

$$Y_{\underset{\sim}{k}} = \sum_{\underset{\sim}{i}\in Z^d} c_{i_1}^{(1)}\cdots c_{i_d}^{(d)}\xi_{\underset{\sim}{k}-\underset{\sim}{i}}\ ,\quad \underset{\sim}{k}=(k_1,\cdots,k_d),$$

and consider the partial sum process

$$D_n(\underset{\sim}{\tau}) = \sum_{\underset{\sim}{k};\underset{\sim}{1}\le\underset{\sim}{k}\le n\underset{\sim}{\tau}} Y_{\underset{\sim}{k}}, \qquad \underset{\sim}{\tau}=(\tau_1,\cdots,\tau_d)\in R_+^d.$$

where $\underset{\sim}{1}\le\underset{\sim}{k}\le n\underset{\sim}{\tau}$ means that $1\le k_i\le n\tau_i$ for all $i(1\le i\le d)$.

THEOREM 4.1. *Under all above conditions,*

(4.2)
$$\frac{1}{\psi(n^d)\psi_1(n)\cdots\psi_d(n)} D_n(\underset{\sim}{\tau}) \longrightarrow_{f.d.} \Delta(\underset{\sim}{\tau}), \quad (n\to\infty),$$

where $\longrightarrow_{f.d.}$ *denotes the convergence of all finite-dimensional distributions.*

Proof. We first note that $D_n(\underset{\sim}{\tau})$ may be rewritten as follows;

$$D_n(\underset{\sim}{\tau}) = \sum_{\underset{\sim}{i}} \left(\prod_{q=1}^{d} \sum_{k=1}^{[n\tau_q]} c_{k-i_q}^{(q)}\right)\xi_{\underset{\sim}{i}} = \sum_{\underset{\sim}{i}} \left(\prod_{q=1}^{d} \sum_{k=1-i_q}^{[n\tau_q]-i_q} c_k^{(q)}\right)\xi_{\underset{\sim}{i}}\ .$$

Therefore, the left-hand side of (4.2) may be expressed as $X_n(\prod_{q=1}^d f_n^{(q)}(\tau_q;u_q))$, where

$$f_n^{(q)}(\tau;u) = \frac{1}{\psi_q(n)} \sum_{k=1-[nu]}^{[n\tau]-[nu]} c_k^{(q)}, \quad \tau>0,\ u\in R,\ (1\le q\le d).$$

However, we have shown in the proofs of Theorems 5.1 and 5.2 of [8] that, for every q $(1\le q\le d)$ and every fixed $\tau>0$,

(C1) $f_n^{(q)}(\tau;u)$ converges continuously to $f_{a_q,b_q,\gamma_q}(\tau;u)$ as $n\to\infty$,

(C2) for every $T>0$, there exists a $\beta(>\alpha)$ such that

$$\sup_n \int_{|u|\le T} |f_n^{(q)}(\tau;u)|^{\beta}\, d\rho_n(u) < \infty, \quad \rho_n(u) = \frac{[nu]}{n}$$

and

(C3) there exists an $\varepsilon_0(>0)$ such that

$$\lim_{T\to\infty}\ \limsup_{n\to\infty} \int_{|u|>T} \{|f_n^{(q)}(\tau;u)|^{\alpha-\varepsilon_0}+|f_n^{(q)}(\tau;u)|^{\alpha+\varepsilon_0}\}\mu_n(du) = 0.$$

Therefore, if we fix $\underset{\sim}{\tau}\in R_+^d$ and apply Theorem 3.3 to $f_n(\underset{\sim}{u}) = \prod_{q=1}^d f_n^{(q)}(\tau_q;u_q)$ and $f(\underset{\sim}{u}) = \prod_{q=1}^d f_{a_q,b_q,\gamma_q}(\tau_q;u_q)$, we have

$$X_n(\prod_{q=1}^{d} f_n^{(q)}(\tau_q;u_q)) \xrightarrow{\ \mathcal{L}\ } X(\prod_{q=1}^{d} f_{a_q,b_q,\gamma_q}(\tau_q;u_q)), \quad (n\to\infty),$$

which may be restated as

$$\frac{1}{\psi(n^d)\psi_1(n)\cdots\psi_d(n)} D_n(\underset{\sim}{t}) \xrightarrow{\mathcal{L}} \Delta(\underset{\sim}{t}), \quad (n \to \infty),$$

for every $\underset{\sim}{t} \in R_+^d$. By the Cramér-Wold method we also have the convergence of all finite-dimensional distributions, which completes the proof of Theorem 4.1.

REFERENCES

[1] Astrauskas, A. (1983): Limit theorems for linearly generated random variables. Lithuanian Mat. J. 23(2), 127-134.

[2] Avram, F. and Taqqu, M. S.(1986): Weak convergence of moving averages with infinite variance. Dependence in Probability and Statistics (ed. by Eberlein and Taqqu), Birkhäuser, Boston. 399-415.

[3] Billingsley, P.(1968): Convergence of Probability Measures. Wiley, New York.

[4] Durrett, R. and Resnick, S. I.(1978): Functional limit theorems for dependent variables, Ann. Probab. 6, 829-846.

[5] Ikeda, N. and Watanabe, S. (1981): Stochasitic Differential Equations and Diffusion Processes, North-Holland, Amsterdam/ Kodansha, Tokyo.

[6] Jagers, P.(1974): Aspects of random measures and point processes, Adv. Probab. Related Topics, 3, Marcel Dekker, New York, 179-239.

[7] Kasahara, Y. and Maejima, M. (1986): Functional limit theorems for weighted sums of i.i.d. random variables. Probab. Th. Rel. Fields 72,161-183.

[8] Kasahara, Y. and Maejima, M.: Weighted sums of i.i.d. random variables attracted to integrals of stable processes. (Preprint)

[9] Kasahara, Y. and Watanabe, S.: Limit theorems for point processes and their functionals. J. Math. Soc. Japan 38,543-574.

[10] Maejima, M. (1983): On a class of self-similar processes. Z. Wahrsch. Verw. Geb. 62, 235-245.

[11] Taqqu, M.S. and Wolpert, R.L. (1983): Infinite variance self-similar processes subordinate to a Poisson measure. Z. Wahrsch. verw. Geb. 62, 53-72.

IMAGE DES POINTS CRITIQUES D'UNE APPLICATION REGULIERE

Paul MALLIAVIN
10, rue Saint-Louis-en-l'Ile. 75004 Paris (France)

Y. KATZNELSON
Mathematics, Hebrew University. Jerusalem (Etat d'Israël)

Etant donné un espace gaussien X, muni d'une mesure gaussienne μ, de dimension finie ou infinie, on note par $\mathscr{W}^{\infty}(X)$ l'intersection de tous les espaces de Sobolev sur X. Etant donné $g \in \mathscr{W}^{\infty}(X; \mathbb{R}^n)$ on considère le déterminant

$$\det(g) = \det(\nabla g^i \cdot \nabla g^j) .$$

L'ensemble des points critiques est défini

$$\mathscr{C}(g) = \{x \; ; \; \det(g)(x) = 0\}$$

On sait que $(g)_* ((1 - \mathbb{1}_{\mathscr{C}})\mu)$ a une densité relativement à la mesure de Lebesgue de $\mathbb{R}^n$. On se propose d'étudier ici la partie résiduelle

$$(g)_* (\mathbb{1}_{\mathscr{C}}\mu) = \nu_{\mathscr{C}} .$$

On a les théorèmes suivants :

THÉORÈME 1. - <u>Il existe</u> $g \in \mathscr{W}^{\infty}(X; \mathbb{R}^n)$ <u>tel que notant</u>

$$\mathscr{C}_0 = \{x \; ; \; \nabla g^i(x) = 0 \quad , \; i \in [1,n]\}$$

<u>alors</u>

$\nu_{\mathscr{C}_0}$ <u>possède une densité relativement à la mesure de Lebesgue de</u> $\mathbb{R}^n$, <u>densité qui de plus est une fonction indéfiniment différentiable</u>.

Appelons rang de g en x_0 le rang de la matrice $\nabla g_i . \nabla g_j$. Soit

$$\mathscr{C}_r = \{x \; ; \; (\text{rang de } g \text{ en } x) = r\} .$$

THÉORÈME 2. - <u>Notons par</u> $\mathscr{C}_r'$ <u>l'intérieur fin de</u> $\mathscr{C}_r$, <u>alors</u>

$\nu_{\mathscr{C}_r'}$ <u>est porté par un ensemble de dimension de Hausdorff</u> $n - r$,

$\nu_{\mathscr{C}_0'}$ <u>est une mesure discrète</u>.

THÉORÈME 3. - Supposons dimension $(X) < \infty$, alors $\nu_{\mathscr{C}_r}$ est porté par un ensemble de dimension de Hausdorff $(n - r)$.

Le théorème 1 a été démontré dans [4] .

1. Preuve du théorème 2.

Nous donnerons d'abord une preuve dans le cas où g est à valeurs scalaires.

Nous montrerons alors

THÉORÈME 4. - Soit $g \in \mathscr{W}(X)$, soit

$$\mathscr{C} = \{x \ ; \ (\nabla g)(x) = 0\} .$$

Notons par $\mathcal{L}$ l'opérateur de Ornstein Uhlenbeck sur X , alors

$$\mathcal{L} \nabla g(x) = 0 \qquad x \in \mathscr{C}$$

$$\mathcal{L} g(x) = 0 \qquad x \in \mathscr{C}$$

et plus généralement l'algèbre d'opérateurs engendrée par les opérateurs $\mathcal{L}$ et ∇ annulent g sur $\mathscr{C}$.

Remarque. On peut considérer ce théorème comme l'analogue du fait en dimension finie qu'une fonction C^∞ dont le gradient s'annule sur un ensemble de mesure positive, a toutes ses dérivées d'ordre > 1 nulles sur cet ensemble. Ce fait va être développé dans le lemme suivant :

1.1. LEMME. - Soit G un espace de Hilbert séparable abstrait. Soit $\varphi \in \mathscr{W}(X,G)$.

Posons

$$K = \{x \ ; \ \varphi(x) = 0\}.$$

Supposons $\mu(K) > 0$, alors

$$\mathcal{L} \varphi \ , \nabla \varphi \ , \mathcal{L} \nabla \varphi \ , \ \dots \ \nabla \mathcal{L} \varphi \ , \ \nabla^2 \varphi \dots$$

sont nuls sur K .

Preuve. Supposons $\nabla \varphi(x) \not\equiv 0$ p.p. K .

Alors il existerait $K_1 \subset K$ et il existerait $h_o \in G$ tel que

$$(\nabla \varphi | h_o)(x) > \beta > 0 \qquad \forall x \in K_1 .$$

Considérons la fonction scalaire :

$$u(x) = (\varphi(x) | h_o) .$$

Notons par $x_\omega(t)$ le processus d'Ornstein Uhlenbeck sur X. Prenons des redéfinitions convenables des fonctions intervenant de telle sorte qu'elles soient continues p.s. sur les trajectoires du processus O.U. Alors en remplaçant K_1 par son adhérence fine K_1^f qui sera contenue dans K, on aura que

$$P_\omega = \{t \in [0,1] , \ x_\omega(t) \in K_1^f\}$$

est une partie fermée. Posons

$$e_\omega(t) = \sup \{ \lambda \in P_\omega , \lambda \leq t \}$$

$$\delta_\omega(t) = t - e_\omega(t) .$$

Alors, par la formule de Ito, il existe un brownien abstrait , tel que

$$u(x_\omega(t)) - u(x_\omega(e_\omega(t))) = \int_{e(t)}^{t} \|\nabla u\| db + \int_{e(t)}^{t} \mathcal{L}u \, dt .$$

Nous avons une minoration sur P_ω de $\|\nabla u\|$. Par suite il existe une partie ouverte O_ω de $[0,1]$, $O_\omega \supset P_\omega$ telle que $|\nabla u(x_o(t))| > \frac{\beta}{2}$ si $t \in O_\omega$.

Lorsque $t \in O_\omega$, on peut subordonner le drift $\mathcal{L}u$ à db par la formule de Girsanov. Par suite $t \in O_\omega$, on obtient que u est donné par un changement de temps d'un brownien , soit b_1. Plus précisément on note

$$\varphi_\omega(t) = \int^{t} \|\nabla u\|^2 (x_\omega(\xi)) d\xi$$

et par Θ_ω la fonction inverse de φ_ω.

Alors, à une équivalence de mesure à la Girsanov près, on a

$$u(x_\omega(t)) = b_1(\Theta_\omega(t)), \quad t \in O_\omega .$$

On sait que le temps local de b_1 a pour dimension de Haursdorff $\frac{1}{2}$. Comme Θ_ω est une transformation absolument continue ceci entraîne

dimension Hausdorff de $P_\omega = \frac{1}{2}$ p.s.

D'autre part comme $\mu(K_1) > 0$

Prob{ mesure de Lebesgue $(P_\omega) > 0\} > 0$,

contradiction qui démontre

$$\nabla \varphi = 0 \quad , \text{ p.p. sur } K .$$

Maintenant posons

$$\nabla \varphi = v .$$

Alors $v \in \mathscr{W}(X ; G \otimes H)$, v est nul sur K par suite

$$(\nabla v)(x) = 0 \text{ , p.p. sur } K .$$

En continuant ainsi de suite on obtient que $\nabla^2 u$, $\nabla^3 u$... sont nuls p.p. sur K .

En dimension finie l'opérateur de Ornstein-Uhlenbeck s'écrit à l'aide des dérivées premières et secondes. Nous allons faire de même en dimension infinie en utilisant un passage à la limite.

Soit $e_1,\dots,e_n$, ... une base orthonormée de H. Soit V_n le sous-espace vectoriel de H engendré par $e_1,\dots,e_n$. Posons

$$(\mathcal{L}_n \varphi)(x) = \text{Trace } \nabla^2 \varphi|_{V_n} - P_{V_n} \nabla\varphi(x) .$$

Alors on a

$$\lim(\mathcal{L}_n \varphi)(x) = \mathcal{L}\varphi(x) \quad \text{p.p.}$$

$$\mathcal{L}_n \varphi(x) = 0 \text{ , } x \in K \quad \text{p.p.}$$

Par suite $(\mathcal{L}\varphi)(x) = 0$. ∎

1.2. Preuve du théorème 4.

On pose

$$\varphi = \nabla g .$$

On déduit du lemme 4.1. que $\nabla\varphi$, $\mathcal{L}\,\nabla\varphi$ etc. sont tous nuls. On obtient

$$\mathcal{L}_n g = \operatorname*{trace}_{V_n} \nabla^2 g - (P_{V_n} x)\ \nabla g = 0$$

d'où $\mathcal{L} g = 0$.

1.3. Preuve du théorème 2 lorsque g est scalaire .

Comme $\mathscr{C}'$ est un ouvert fin , on a

$$P'_\omega = \{ t \ ; \ x_\omega(t) \in \mathscr{C}' \} \text{ est un ouvert de } [0, 1] .$$

Soit $[\alpha\beta] \subset P'_\omega$. Comme sur α, β tous les invariants de Ito de $g(x_\omega(t))$ sont nuls on en déduit, par la formule de Ito que $g(x_\omega(t)) = Cte$, $t \in [\alpha,\beta]$. Ainsi $g(x_\omega(t))$ est constant sur chaque composante connexe fine de $\mathscr{C}'$. Comme chaque composante connexe a une mesure positive, il existe au plus un nombre dénombrable de telles composantes et finalement $g(\mathscr{C}')$ est un ensemble dénombrable.

1.4. Preuve du théorème 2.

Preuve. En choisissant les indices on se ramène au cas où $\nabla g_1, \ldots, \nabla g_r$ sont linéairement indépendants sur $\mathscr{C}'$. Notons $\hat{g}(x) = (g_1(x), \ldots, g_r(x))$, alors on sait que $(\hat{g})_*(1\!\!1_{\mathscr{C}'}\mu) = \rho$ est absolument continue relativement à la mesure de Lebesgue. Fixons $\xi \in R^p$ et considérons la sous-variété [1] de X définie par

$$V_\xi = \hat{g}^{-1}(\xi) \quad .$$

(Nous utilisons maintenant des redéfinitions des fonctions de $\mathscr{W}$ à l'extérieur d'ensembles minces).

En utilisant le théorème des fonctions implicites [1] on obtient qu'il existe le sous-espace gaussien Y , de codimension r dans X et une décomposition en somme directe $X = Y \oplus Z$

$$x = (y \, , \, z)$$

tel que sur un ouvert fin de Y le changement de coordonnées

$$(y,z) \longrightarrow (y,\hat{g}(y,z))$$

soit un difféomorphisme local (ce difféomorphisme peut être considéré comme un difféomorphisme local de $\mathbb{R}^r$, dépendant finement continûment du paramètre y , soit $\hat{g}_y$).On note par ψ_y le difféomorphisme local réciproque de $\hat{g}_y$.

On peut ainsi identifier localement V_ξ à Y. La topologie fine sur V_ξ sera celle définie par Y. Cette topologie ne dépend pas du choix du supplémentaire Y.

LEMME. - <u>Presque partout</u> ρ, $V_\xi \cap \mathscr{C}'$ <u>est un ouvert fin de</u> V_ξ.

<u>Preuve</u>. Le processus d'Ornstein Uhlenbeck s'écrit

$$x_\omega(t) = (y_{\omega_1}(t), z_{\omega_2}(t)).$$

les deux processus dans le second membre étant indépendants. Par suite un ouvert fin $\mathscr{C}'$ de X se projette sur Y suivant un ouvert fin $\mathscr{C}'_1$.

En effet dire que $\mathscr{C}'$ est un ouvert fin est équivalent à dire $x_\omega^{-1}(\mathscr{C}')$ est un ouvert de $[0, 1]$, p.s.

Donc si $x_\omega(0) \in \mathscr{C}'$, il existe $\varepsilon(\omega) > 0$ tel que $x_\omega(t) \in \mathscr{C}'$ pour $|t| < \varepsilon(\omega)$ a fortiori $y_{\omega'}(t) \in \Pi_Y(\mathscr{C}')$ si $|t| < \varepsilon(\omega)$.

D'autre part, d'après le théorème des fonctions implicites donne, que localement, ξ_0 fixé,

$$\Pi_Y(\mathscr{C}' \cap V_{\xi_0}) = \Pi_Y(\mathscr{C}').$$

Définissons des fonctions $\mathscr{W}_{loc,fin}(Y)$ en posant

$$\gamma_p(y) = g_p(\psi_y(\xi_0)) \qquad p > r.$$

Alors comme sur V_{ξ_0}, ∇g_p est combinaison linéaire des ∇g_s, $s < r$, on en déduit que

$$\nabla \gamma_p = 0 \qquad \text{sur} \qquad \Pi_Y(\mathscr{C}').$$

Par suite en utilisant le résultat scalaire, g_p prend sur $\mathscr{C}' \cap V_{\xi_0}$ un nombre dénombrable de valeurs et $(g)_* 1\!\!1_{\mathscr{C}'} \mu$ est porté par un ensemble $R^r \times D$ où D est une partie dénombrable de R^{n-r}. ∎

2. <u>Preuve du théorème 3</u>.

Cette démonstration dépend de considérations d'<u>entropie</u> propre à la dimension finie.

D'après le théorème d'immersion de Sobolev la fonction g est une fonction C^∞. Le théorème usuel des fonctions implicites est disponible. Par suite, utilisant l'approche du paragraphe précédent, il suffit de démontrer le théorème lorsque g est scalaire plus précisément.

2.0. Soit $g \in \mathscr{W}^0(X;\mathbb{R})$, alors $\nu_{\mathscr{C}}$ est porté par un ensemble de mesure de Hausdorff nulle.

Munissons X d'une métrique euclidienne d, alors

2.1. LEMME. - Pour tout $\varepsilon > 0$, et pour tout $p > 0$, on peut trouver $\mathscr{C}_1 \subset \mathscr{C}$, tel que $\mu(\mathscr{C}_1) > (1-\varepsilon)\,\mu(\mathscr{C})$ et tel que

$$|g(x) - g(x_0)| < c\,d^p(x,x_0) \quad \text{pour tout } x_0 \in \mathscr{C} \quad \text{(c étant une constante convenable)}.$$

Preuve. L'estimation non uniforme en x_0 résulte de la formule de Taylor et du fait que toutes les dérivées de g s'annulent en x_0. L'estimation uniforme s'obtient en utilisant les idées de Egoroff et Lusin.

2.2. LEMME. - Fixons $c > \ell = \dim X$, alors on peut trouver $\mathscr{C}_2$, $\mathscr{C}_2 \subset \mathscr{C}_1$ tel que $\mu(S_2) > (1-\varepsilon)\mu(S_1)$, et tel qu'il existe r_0 tel que pour tout $r > r_0$, on puisse recouvrir $\mathscr{C}_2$ avec 2^{cr} cubes de côté 2^{-r}.

Preuve. En diminuant la mesure de $\mathscr{C}_1$, on peut se ramener au cas où $\mathscr{C}_2$ est compac Alors la mesure μ est équivalente à la mesure de Lebesgue au voisinage de $\mathscr{C}_2$. Introduisons la fonction maximale

$$\psi(x_0) = \sup_r \frac{1}{2^{-r\ell}} \int_{\Delta_r(x_0)} 1\!\!1_{\mathscr{C}_1}\, dx$$

où $\Delta_r(x_0)$ dénote le cube centré x_0 et de côté 2^{-r} et où $\ell = \dim X$.

Alors d'après le théorème de dérivation $\psi(x_0) = 1\!\!1_{\mathscr{C}_1}(x_0)$ p.p. Utilisant le théorème d'Egoroff on peut trouver $\mathscr{C}_2 \subset \mathscr{C}_1$ tel que cette convergence soit uniforme. En particulier $\int_{\Delta_r(x_0)} 1\!\!1_{\mathscr{C}_1} \geq 2^{-r\ell}\gamma$ si $r > r_0$ et $x \in \mathscr{C}_2$ où γ est une constante fixée, $> 1 - 2^{-\ell-1}$.

Fixons un système de coordonnée et considérons un maillage par les cubes dont les sommets sont de la forme $j2^{-r}$. Ayant fixé r, soit P_r la famille de ces cubes rencontrant $\mathscr{C}_2$.

Soit $\Delta' \in \Gamma_r$, $x_o \in \Delta' \cap \mathscr{C}_2$. Alors $\Delta_{r-1}(x_o) \supset \Delta'$. Par suite

$$\int_{\Delta'} 1\!\!1_{\mathscr{C}_1} > \frac{1}{2} \operatorname{vol}(\Delta')$$

d'où

$$\Sigma \operatorname{vol}(\Delta') < 2 \operatorname{vol}(\mathscr{C}_1) . \quad \blacksquare$$

2.3. Preuve de 2.0.

Il résulte de 2.1. et de 2.2. que l'on peut recouvrir $g(\mathscr{C}_2)$ par 2^{cr} intervalles chacun de longueur $< 2^{-pr}$. Par suite la

$$\dim_{\text{Hausdorff}}(g(\mathscr{C}_2)) \leq \frac{c}{p} .$$

Comme on peut prendre p arbitrairement grand

$$\dim_{\text{Hausdorff}}(g(\mathscr{C}_2)) = 0 . \quad \blacksquare$$

Bibliographie

1] H.AIRAULT et P.MALLIAVIN. - Intégration géométrique sur l'espace de Wiener. Bulletin Sciences Math., 1988, p. 13-55.

2] GETZLER. - Index theorem on the Wiener space. Journal Functional Analysis, 1985 et 1987.

3] S.KUSUOKA. - Index theorem on the Wiener space and new infinitesimal geometric invariants Colloquium Franco-Japonais, Juin 1987.

4] Y.KATZNELSON et P.MALLIAVIN. - Un contre exemple pour le théorème de Sard en dimension infinie. Comptes-Rendus, décembre 1987.

5] Y.YOMDIM. - Conter examples to the Sard's theorem in infinite dimension, preprint.

DEGREE THEOREM IN CERTAIN WIENER RIEMANNIAN MANIFOLDS

Shigeo KUSUOKA
RIMS. University of Kyoto, Kyoto, 606 (Japan)

In this paper, we consider the degree theorem in some special Wiener Riemannian manifolds. This is an extension of Getzler [3] and our previous work [6]. We also show in Section 3 that Gauss-Bonnet-Chern formula follows from our degree theorem, and show the formula (3.6) which is an analogue of the formula in Bismut [1] and is cojectured by supersymmetry argument.

1. Sobolev spaces.

In this section, we introduce Sobolev spaces over special Wiener Riemannian manifolds by the same way as Sobolev spaces over Wiener spaces (c.f. Kree[5], Watanabe[11]).

Let (μ, H, B) be a abstract Wiener space, i.e., B is a separable real Banach (or Frechet) space, H is a separable real Hilbert space continuous and densely embedded in B, and μ is a Gaussian probability measure on B satisfying

$$\int_B \exp(i\,{}_B\langle z,u\rangle_{B^*})\mu(dz) = \exp(-\frac{1}{2}\|u\|_H^2), \quad u \in B^* \subset H,$$

where we identify the dual space H^* of H with H.

Let M and N be finite dimensional Riemannian manifolds and E be a Hilbertian vector bundle over N with Hilbertian connection. We assume that a fiber space of E is a separable Hilbert space.

(1.1) Definition. We say that $F: M\times B \to E$ is a $\mathcal{H}$-C^0 map, if $F(.,z+*): M\times H \to E$ is continuous for each $z \in B$.

(1.2) Definition. We say that $F: M\times B \to E$ be a compact $\mathcal{H}$-C^0 map if for $F(x_n, z+h_n) \to F(x,z)$ in E for any $z \in B$, $x \in M$, $\{x_n\}_1^\infty \subset M$ and $(h_n)_1^\infty \subset H$ such that $x_n \to x$ in M and $h_n \to 0$ weakly in H as $n \to \infty$.

Since a Hilbertian connection on E is given, we can identify the tangent space $T_e(E)$ of E at $e \in E$ with $T_{\pi(e)}(N) \oplus E_{\pi(e)}$, where $\pi:E\to N$ is a natural projection. ($T_{\pi(e)}(N)$ is the horizontal part and $E_{\pi(e)}$ is the vertical part). We denote by $\mathscr{H}_M(E)$ a Hilbert vector bundle over M×N such that the fiber space $\mathscr{H}_M(E)_{(x,y)}$ at $(x,y) \in M\times N$ is the space of Hilbert-Schmidt linear operators from $T_x(M)\oplus H$ into $T_y(N)\oplus E_y$. $\mathscr{H}_M(E)$ has a natural induced Hilbertian connection.

(1.3)<u>Definition.</u> We say that a map $F:M\times B\to E$ is an $\mathscr{H}$-C^1 map if

(0) F is an $\mathscr{H}$-C^0 map,

(1) there is an $\mathscr{H}$-C^0 map $DF:M\times B\to\mathscr{H}_M(E)$ such that

(i) $\pi^{(1)}DF(x,z) = (x,\pi(F(x,z))) \in M\times N$ and

(ii) $DF(x,z)(v,h) = \frac{d}{dt} F(\exp_x(tv),z+th)|_{t=0}$

$$\in T_{F(x,z)}(E) \simeq T_{\pi(F(x,z))}(N)\oplus E_{\pi(F(x,z))}$$

for each $(x,z) \in M\times B$. Here $\pi:E\to N$ and $\pi^{(1)}:\mathscr{H}_M(E)\to M\times N$ are natural projections.

(1.4)<u>Definition.</u> We say that $F:M\times B\to E$ is a compact $\mathscr{H}$-C^1 map if

(1) $F:M\times B\to E$ is an $\mathscr{H}$-C^1 map and

(2) $F:M\times B\to E$ and $DF:M\times B\to\mathscr{H}_M(E)$ are compact $\mathscr{H}$-C^0 maps.

Also, We define (compact) $\mathscr{H}$-C^n maps inductively in the following.

(1.5)<u>Definition</u>. We say that $F:M\times B\to E$ is a (compact) $\mathscr{H}$-C^n map, $n \geq 2$, if

(1) $F:M\times B\to E$ is a (compact) $\mathscr{H}$-C^1 map and

(2) $DF:M\times B\to\mathscr{H}_M(E)$ is a (compact) $\mathscr{H}$-C^{n-1} map.

For each $\mathscr{H}$-C^n map $F:M\times B\to E$, we have maps $D^kF:M\times B\to\mathscr{H}_M{}^k(E)$, $k = 1,\ldots,n$, where $\mathscr{H}_M{}^k(E) = \underbrace{\mathscr{H}_M(\mathscr{H}_M(\ldots(\mathscr{H}_M}_{k\text{-times}}(E))\ldots))$. As a convention we write D^0F and $\mathscr{H}_M{}^0(E)$ for F and E respectively.

Let $\tilde{\mathscr{E}}^n(M;E) = \{F:M\times B\to E;$ F is an $\mathscr{H}$-C^n maps and

$$\sup_{x\in K\ z\in B} \|D^kF(x,z)\|_{\mathscr{H}_M{}^k(E)} < \infty$$

for any compact set K in M, $0\leq k\leq n$ },

for each $n \geqq 0$.

For each compact subset K in M, $n \geqq 0$ and $1 < p < \infty$, let

$$\|F\|_{\mathcal{E}^n_{p,K}} = (\int_{K\times B} \sum_{k=0}^{n} \|D^k F(x,z)\|^p_{\mathcal{H}_M^k(E)}\, m(dx)\otimes\mu(dz))^{1/p},\quad F\in\tilde{\mathcal{E}}(M;E).$$

Then for each $n \geqq 0$ and $1 < p < \infty$, $\{\|\ \|_{\mathcal{E}^n_{p,K}}$; K compact subsets in M$\}$ is a family of semi-norms on $\tilde{\mathcal{E}}(M;E)$. Let $\mathcal{E}^n_p(M;E)$ be the completion with respect to this family of semi-norms.

For each compact subset K in M, $n \geqq 0$ and $1 < p < \infty$, let

$\mathcal{D}^n_{p,K}(M;E) = \{F\in\mathcal{E}^n_p(M;E);\ F(x,z) = 0$ for $m\otimes\mu$-a.e.$(x,z) \in K^c\times B\}$. Then $\mathcal{D}^n_{p,K}(M;E)$ is a closed subspace of $\mathcal{E}^n_p(M;E)$. We define the spaces $\mathcal{E}^\infty_p(M;E)$, $\mathcal{D}^\infty_p(M;E)$, $1+ \leqq p \leqq \infty-$, by

$$\mathcal{E}^\infty_p(M;E) = \bigcap_{n\geqq 1} \mathcal{E}^n_p(M;E),\quad \mathcal{D}^\infty_p(M;E) = \varinjlim_{K} \bigcap_{n\geqq 1} \mathcal{D}^n_{p,K}(M;E),\quad 1 < p < \infty,$$

$$\mathcal{E}^\infty_{1+}(M;E) = \varinjlim_{p\to 1} \mathcal{E}^\infty_p(M;E),\quad \mathcal{D}^\infty_{1+}(M;E) = \varinjlim_{p\to 1} \mathcal{D}^\infty_p(M;E),$$

$$\mathcal{E}^\infty_{\infty-}(M;E) = \bigcap_{1<p<\infty} \mathcal{E}^\infty_p(M;E) \text{ and } \mathcal{D}^\infty_{\infty-}(M;E) = \varinjlim_{K} \bigcap_{1<p<\infty} \bigcap_{n\geqq 1} \mathcal{D}^n_{p,K}(M;E)\ .$$

Then we can think of the dual spaces of these spaces and also of the Malliavin calculus. Since M and N are finite dimensional manifolds, the Malliavin calculus in this case is quite similar to the usual Malliavin calculus. Moreover, we will not use the Malliavin calculus in this paper so hardly. Thus we stop disscussing on the formalism of the Mallivin calculus.

2. Manifold and Degree.

Let M be a C^∞ Riemannian manifold of dimension d and let $F:M\times B\to M$ be a compact $\mathcal{H}$-C^3 map. We impose the following condition.

(A-1) There is a smooth function $r:M \to (0,\infty)$ such that

(1) $\exp_x:\{v\in T_x(M);\|v\|_{T_x(M)} < 4r(x)\} \to M$ is a diffeomorphism for any $x \in M$, and

(2) $$\sup_{x\in K} \int_{\{z\in B;d(x,F(x,z)<4r(x))\}} \det{}_{TM_x}(D_1F(x,z)D_1F(x,z)^*)^{-p}\mu(dz) < \infty$$

for any compact subset K in M and $1 < p < \infty$.

Here $D_0F(x,z) \in Hom(T_x(M);T_{F(x,z)}(M))$ and $D_1F(x,z) \in Hom(H;T_{F(x,z)}(M))$ $(x,z) \in M\times B$ are given by $D_0F(x,z)(v) = DF(x,z)(0,v)$ and $D_1F(x,z)(h) = DF(x,z)(0,h)$, $(x,z) \in M\times B$, $v \in T_x(M)$, $h \in H$.

Take a $\psi \in C_0^\infty(\mathbb{R};\mathbb{R})$ such that $\psi(t) = 1$, $|t| < \frac{4}{3}$ and $\psi(t) = 0$, $|t| > \frac{5}{3}$, and fix it. Let

$$\rho_s(x,y) = \psi(\frac{d(x,y)}{r(x)})\cdot(\frac{1}{2\pi s})^{d/2}\exp(-\frac{d(x,y)^2}{2s}), \quad s > 0,\ x,y \in M.$$

Then $\rho_s(x,y) \to \delta(x,y)$ in $\mathcal{D}'(M\times M)$ as $s \downarrow 0$.

Define a Radon measure $\tilde{\nu}$ on $M\times B$ by

$$\begin{aligned}\tilde{\nu}(dx\otimes dz) &= \delta(x,F(x,z))\ m(dx)\otimes\mu(dz)\\ &= \text{weak-}\lim_{s\downarrow 0}\ \rho_s(x,F(x,z))\ m(dx)\otimes\mu(dz)\ .\end{aligned}$$

The existance of such a measure $\tilde{\nu}$ is guarenteed by Sugita [9] and Malliavin [8].

Let us define $\tilde{\iota}(x,y):T_x(M)\to T_y(M)$, $x,y \in M$, $d(x,y) < 2r(x)$, as follows: for each u, $v \in T_x(M)$ with $\|u\|_x = 1$, $\tilde{\iota}(x,\exp_x(tu))v$ is a parallel displacement along the curve $\exp_x(tu)$, $t \in [0,2r(x)]$. We define $\iota(x,y):T_x(M)\to T_y(M)$, $x,y \in M$, by $\iota(x,y) = \psi(\frac{d(x,y)}{r(x)})\tilde{\iota}(x,y)$. We define a Radon measure ν on $M\times B$ by

$$\begin{aligned}&\nu(dx\otimes dz)\\ &= \det_{T_{F(x,z)}(M)}((\iota(x,F(x,z))-DF(x,z))(\iota(x,F(x,z))-DF(x,z))^*)^{1/2}\\ &\qquad\qquad\qquad\qquad\qquad\qquad\qquad\qquad\tilde{\nu}(dx\otimes dz)\ .\end{aligned}$$

The measure ν is considered 'the induced measure on the submanifold $\{(x,z)\in M\times B; F(x,z)=x\}$ in $M\times B$'.

Let us define $E_0(x,z) \in Hom(T_x(M)\oplus H;T_{F(x,z)}(M))$, $(x,w) \in M\times B$ by

$$E_0(x,z) = \iota(x,F(x,z)) - DF(x,z)\ .$$

Also let us define $E(x,z) \in Hom(T_x(M)\oplus H;T_x(M)\oplus H)$ and $S(x,z) \in Hom(T_x(M)\oplus H;T_x(M))$, $(x,z) \in M\times B$, by

$$E(x,z) = I_{TM_x\oplus H} - E_0(x,z)(E_0(x,z)E_0(x,z)^*)^{-1}E_0(x,z)$$ and

$S(x,z) = P_{TM_x} - \iota(F(x,z),x)DF(x,z)$, where P_{TM_x} is the orthogonal projection in $T_x(M)\oplus H$ onto $T_x(M)$. Then it is easy to see that $S(x,z)E(x,z) = 0$ ν-a.e.(x,z).

(2.1)<u>Proposition</u>. For any $u \in \mathcal{E}^{\infty}_{\infty-}(M;TM\oplus H)$ and $f \in \mathcal{D}^{\infty}_{\infty-}(M;R)$

$$\int_{M\times B} f(x,z)(E(x,z)u(x,z),D(\delta(x,F(x,z))))_{T_x(M)\oplus H}\, m(dx)\otimes\mu(dz) = 0.$$

<u>Proof</u>. Let $x \in M$ and $a,b \in T_x(M)$. Let $\gamma_0:(-\varepsilon,\varepsilon)\to M$ and $\gamma:(-\varepsilon,\varepsilon)\times(-\varepsilon,\varepsilon)\to M$ be given by $\gamma_0(t) = \exp_x(ta)$ and $\gamma(t,s) = \exp_{\gamma_0(t)}(s\iota(x,\gamma_0(t)))$ for sufficiently small $\varepsilon > 0$. Then

$d(\gamma(t,0),\gamma(t,s)) = |s|\,\|b\|_x$ and so $\frac{d}{dt}\, d(\gamma(t,0),\gamma(t,s)) = 0$. Note that $\frac{d}{dt}\,\gamma(t,s)|_{t=s=0} = a$ and $\frac{D}{ds}\frac{d}{dt}\,\gamma(t,s)|_{t=s=0} = \frac{D}{dt}\frac{d}{ds}\,\gamma(t,s)|_{t=s=0}$
$= \frac{=}{dt}\,\iota(x,\gamma_0(t))|_{t=0} = 0$. Therefore we see that

$$\|\, \frac{d}{dt}\,\gamma(t,s)|_{t=0} - \iota(x,\gamma_0(0,s))a\, \|_{\gamma(0,s)} = O(s^2) \text{ as } s \to 0.$$

This observation shows us that for any compact subset K in M, there is a $C < \infty$ such that

$$|\, \langle\, d\rho_s(x,y),\ (a,\iota(x,y)a)\, \rangle_{(x,y)}| \leqq C\, \|a\|_x\, \frac{d(x,y)^3}{s} \exp(- \frac{d(x,y)^2}{2s})$$

for any $x \in K$, $y \in M$, $a \in T_x(M)$ and $s \in (0,1]$.

Since $(E(x,z)u(x,z),D(\rho_s(x,F(x,z))))_{T_x(M)\oplus H}$

$$= \langle\, d\rho_s\ ,\ (\iota(x,x)E(x,z)u(x,z),DF(x,z)E(x,z)u(x,z))\, \rangle_{(x,F(x,z))},$$

we see that

$$\int_{M\times B} f(x,z)(E(x,z)u(x,z),D(\rho_s(x,F(x,z))))_{T_x(M)\oplus H}\, m(dx)\otimes\mu(dz) \to 0$$

as $s \downarrow 0$. Q.E.D.

(2.2)<u>Definition</u>. For any $F \in \mathcal{E}^{\infty}(M;TM\oplus H)$, we define $\partial\mathrm{iv}\, F \in \mathcal{E}^{\infty}(M;R)$ by $(\partial\mathrm{iv}\, F)(x,z) = -\operatorname{div} F_1(\cdot,z)|_{\cdot=x} + \partial F_2(x,\cdot)|_{\cdot=z}$, $(x,z) \in M\times B$, where $F = (F_1,F_2)$, $F_1 \in \mathcal{E}^{\infty}(M;TM)$ and $\mathcal{E}^{\infty}(M;H)$.

Then the following is obvious.

(2.3)<u>Lemma</u>. For any $F \in \mathcal{E}^{\infty}_{\infty-}(M;TM\oplus H)$ and $f \in \mathcal{E}^{\infty}_{\infty-}(M:R)^*$,

$$\int_{M\times B} (F(x,z),Df(x,z))_{TM_x\oplus H}\, m(dx)\otimes\mu(dz) = \int_{M\times B} \partial\mathrm{iv}F(x,z)f(x,z)\, m(dx)\otimes\mu(dz)\ .$$

Now let $\varphi: M\times B\to\mathbf{R}$ be an $\mathcal{H}$-C^0 map satisfying

(A-2) $\varphi \in \mathcal{E}^{\infty}(M;\mathbf{R})$ and $\varphi(x,z) = 0$ or 1 ν-a.e.(x,z)

and $V: M\times B\to H$ be a compact $\mathcal{H}$-C^3 map with $V \in \mathcal{E}^{\infty}_{\infty-}(M;H)$. We think of a 'manifold' $\tilde{M} = \tilde{M}_{F,\varphi} = \{(x,z)\in M\times B;\ F(x,z) = x$ and $\varphi(x,z) = 1\}$ and a map $\Phi: M\times B\to B$ given by $\Phi(x,z) = z + V(x,z)$, and we want to consider $\deg(\Phi|_{\tilde{M}})$, the degree of the map $\Phi|_{\tilde{M}}: \tilde{M}\to B$.

(2.4)<u>Definition</u>. We define a signed measure $(\Phi|_{\tilde{M}})^*\mu$ on $M\times B$ by

$$\begin{aligned}
&(\Phi|_{\tilde{M}})^*\mu(dx\otimes dz)\\
&= \varphi(x,z)\ \det{}_{2\ TM_x\oplus H}((P_H+DV(x,z))(I - P(x,z))+K(x,z))\\
&\qquad \times\exp(-\partial V(x,z) - \tfrac{1}{2}\|V(x,z)\|_H^2 + \mathrm{trace}_{TM_x\oplus H}\ R(x,z))\ \nu(dx\otimes dz).
\end{aligned}$$

Here P_H is the orthogonal projection onto H, I is identity map in $TM_x\oplus H$, $\partial V(x,z) = \partial V(x,\cdot)|_{\cdot=z}$,

$K(x,z) = (S(x,z)S(x,z)^*)^{-1/2}S(x,z)$,

$P(x,z) = K(x,z)K(x,z)^*$ and

$R(x,z) = (P_H + DV(x,z))(I - P(x,z)) + K(x,z) - I - DV(x,z)$.

(2.5)<u>Remark</u>. It is easy to see that

$$\begin{aligned}
&(\Phi|_{\tilde{M}})^*\mu(dx\otimes dz)\\
&= \varphi(x,z)\det{}_{2\ TM_x\oplus H}((P_H+DV(x,z))(I - P(x,z)) + S(x,z))\\
&\qquad \times\exp(-\partial V(x,z) - \tfrac{1}{2}\|V(x,z)\|_H^2 + \mathrm{trace}_{TM_x\oplus H}R_1(x,z))\ \tilde{\nu}(dx\otimes dz),
\end{aligned}$$

where $R_1(x,z) = (P_H+DV(x,z))(I-P(x,z)) + S(x,z) - I - DV(x,z)$.

On the other hand, since

$$\begin{aligned}
&\|R(x,z)\|_{H.S(TM_x\otimes H;TM_x\otimes H)}\\
&\leqq \|DV(x,z)P(x,z)\|_{H.S(TM_x\otimes H;TM_x\otimes H)}\\
&\qquad + \|K(x,z)+(P_H-I)+P_H(I-P(x,z))\|_{H.S(TM_x\otimes H;TM_x\otimes H)}\\
&\leqq \|DV(x,z)\|_{H.S(TM_x\otimes H;TM_x\otimes H)} + 3\dim M,
\end{aligned}$$

for each $\delta > 0$, there is a C_δ such that

$$|(\Phi|_{\tilde{M}})^*\mu(dx\otimes dz)|$$

$$\leq C_\delta\ \varphi(x,z)\ \exp\left(\frac{1+\delta}{2}\ \|DV(x,z)\|_{H.S(TM_x\otimes H;H)} - \partial V(x,z) - \frac{1}{2}\|V(x,z)\|^2\right)$$

$$\nu(dx\otimes dz)\ .$$

Then the following is our main result.

(2.6)<u>Theorem</u>. Assume that

(A-3) for any compact set K in M, there is an $\delta > 0$ such that

$$\int_{K\times B} \exp\left((1+\delta)\{\frac{1+\delta}{2}(1+\|DV(x,z)\|_{HS(TM_x\oplus H;H)})^2 - \partial V(x,z) - \frac{1}{2}\|V(x,z)\|_H^2\right)\ m(dx)\otimes\mu(dz) < \infty, \text{ and}$$

(A-4) $\int_{M\times B} |d(\Phi|_{\tilde{M}})^*\mu| < \infty$ and

there is a sequence $\{g_n\}_{n=1}^{\infty} \subset C_0^{\infty}(M;\mathbf{R})$ such that $0 \leqq g_n \leqq 1$, $g_n \uparrow 1$ and $\int_{M\times B} \|\text{grad } g_n\|_x\ \|((P_H+DV(x,z))(I - P(x,z)) + S(x,z))^{-1}\|_{\text{operator}}$

$$|d(\Phi|_{\tilde{M}})^*\mu| \to 0 \text{ as } n \to \infty.$$

Then we have

(1) $\deg(\Phi|_{\tilde{M}}) \underset{\text{def}}{=} \int_{M\times B} d(\Phi|_{\tilde{M}})^*\mu$ is an integer, and

(2) $$\sum_{\substack{\Phi(x,z)=w\\ F(x,z)=x\\ \varphi(x,z)=1}} \text{sign}(\det_2((P_H+DV(x,z))(I-P(x,z))+K(x,z))) = \deg(\Phi|_{\tilde{M}})$$

for μ-a.e.w.

We will give the sketch of the proof in Section 4.

(2.7)<u>Remark</u>. (1) If the manifold M is compact, the assumption (A-4) follows from the assumption (A-3).

(2) Let us think of the special case where M is compact and $V \equiv 0$. Then $(\Phi|_{\tilde{M}})^*\mu(dx\otimes dz) = \varphi(x,z)\det_{TM_x}(\iota(x,F(x,z))-D_0F(x,z))\tilde{\nu}(dx\otimes dz)$ and

$$\sum_{F(x,z)=x} \varphi(x,z)\ \text{sign}(I_{TM_x} - D_0F(x,z)) = \deg(\Phi|_{\tilde{M}}) \quad \mu\text{-a.e.z.}$$

Moreover, if $\varphi \equiv 1$, then $\deg(\Phi|_{\tilde{M}})$ = Lefschetz number of $F(\cdot,z):M\to M$ for μ-a.e.z.

<u>3. Examples.</u>

Let $\Theta = \{\theta:[0,1]\to\mathbf{R}^d;$ continuous and $\theta(0) = 0\}$ and μ be the standard Wiener measure on Θ and let H be the Cameron-Martin space of

μ. Then (μ,H,Θ) be an abstract Wiener space. Let M be a smooth Riemannian manifold of dimensions N and V_α, $\alpha = 0,1,\ldots,d$, be smooth vector fields on M. We think of a S.D.E.

$$(3.1)\quad \begin{aligned} dX^\varepsilon(t,x;\theta) &= \varepsilon \sum_{\alpha=1}^{d} V_\alpha(X^\varepsilon(t,x;\theta))\circ d\theta^\alpha(t) + \varepsilon^2 V_0(X^\varepsilon(t,x;\theta))dt \\ X^\varepsilon(t,x;\theta) &= x \in M \end{aligned}$$

for each $\varepsilon > 0$. We assume that the S.D.E. (3.1) has a good unique solution. Let $\psi:M\to M$ be a smooth map and let $F = F^\varepsilon:M\times\Theta\to M$ be a measurable map given by $F(x,\theta) = (X(1,\cdot;\theta)^{-1}\psi)(x)$.

Exaple 1. Let M be a compact manifold and $\{V_0,V_1,\ldots,V_d\}$ satisfy the restricted Hörmander condition, i.e., the vector space spaned by $\bigcup_{n=0}^{\infty} \{[V_{i_n},[V_{i_{n-1}},[\ldots[V_{i_1},V_j]\ldots](x);\ j = 1,\ldots,d,\ i_1,\ldots,i_n = 0,1,\ldots,d\ \}$ is equal to $T_x(M)$. Then $F:M\times B\to M$ is a compact $\mathcal{H}$-C^3 map and satisfy the assumption (A.1) and that $F \in \mathcal{D}^\infty_{\infty-}(M;M)$ (c.f. [6]).

(Example 1.1) Let $\varphi \equiv 1$ and $V \equiv 0$, then from Remark(2.6), we see that $\deg(\Phi|_{\tilde{M}})$ = the Lefschetz number of $(X(1,\cdot;\theta)^{-1}\psi):M\to M$ μ-a.e.θ. Since $(X(t,\cdot;\theta)^{-1}\psi)$ is homotopic to ψ, we see that $\deg(\Phi|_{\tilde{M}})$ = the Lefshetz number of ψ. In particular, if $\psi \equiv$ identity, we see that

$$(3.2)\quad \begin{aligned} \chi(M) &= \deg(\Phi|_{\tilde{M}}) \\ &= \int_{M\times\Theta} \det_{TM_x}(I_{TM_x} - X^\varepsilon(1,\cdot;\theta)_{*x})\delta(X^\varepsilon(1,x;\theta),x)m(dx)\otimes\mu(d\theta). \end{aligned}$$

Here $\chi(M)$ denotes the Euler number of M.

If $L = \frac{1}{2}\sum_{\alpha=1}^{d} V_\alpha^{\ 2} + V_0$ is an elliptic operator, by using Schilder-Bismut-Watanabe's expansion formula (c.f. Ikeda-Watanabe [4]) , one can easily compute the limit of the righthand side of (3.2) as $\varepsilon \downarrow 0$ and we get a certain kind of the Gauss-Bonnet-Chern formula.

S. Takanobu pointed out that even if L is not elliptic, as far as $\{V_\alpha;\ \alpha = 0,1,\ldots,d\}$ satisfies the restricted Hörmander condition, one can compute the limit of the righthand side of (3.2) by using his results in [10] and we get a certain kind of the Gauss-Bonnet-Chern formula again.

(Example 3.2) Now let us assume furthermore that the fundamental goup $\pi_1(M)$ is not trivial. Then for each $g \in \pi_1(M)$, one can find a smooth bounded function $\tilde{\varphi}_g: C([0,1];M) \to R$ such that $\tilde{\varphi}_g(w) = 1$ if $w(0) = w(1)$ and $\{w(t);\ t \in [0,1]\}$ is conjugate to g and $\tilde{\varphi}_g(w) = 0$ if $w(0) = w(1)$ and $\{w(t);\ t \in [0,1]\}$ is not conjugate to g. Let $\varphi: M\times\Theta \to R$ be an $\mathcal{H}$-C^0 map given by $\varphi(x,\theta) = \tilde{\varphi}_g(X(\cdot,x;\theta))$ and $V \equiv 0$. Then we have

(3.3) $\deg(\Phi|_{\tilde{M}})$

$$= \int_{M\times\Theta} \det_{TM_x}(I_{TM_x} - X^{\varepsilon}(1,\cdot;\theta)_{*x})\varphi(x,\theta)\delta(X^{\varepsilon}(1,x;\theta),x)m(dx)\otimes\mu(d\theta).$$

However, it is easy to see that the righthand side of (3.3) is continuous in ε and so independent of ε. Also, it is easy to see that if $g \neq e$, the identity, then the limit of the righthand side of (3.3) as $\varepsilon \downarrow 0$ is zero. Combining this with Theorem(2.5)(3), we have

$$(3,4)\quad \deg(\Phi|_{\tilde{M}}) = \begin{cases} 0 & \text{if } g \neq e \\ \chi(M) & \text{if } g = e \end{cases}.$$

(Example 3.3) We think of more special case. Suppose the compact Riemannian manifold M is embeded in R^D isometrically. Then we can identify $T_x(M)$ with a vector subspace of R^D. For each $x \in M$, let P_x denote the orthogonal projection in R^d onto $T_x(M)$. Let $e_\alpha = (0,\ldots,0,\overset{\alpha}{1},0,\ldots,0)$, $\alpha = 1,\ldots,d$, and let $V_0(x) = 0$ and $V_\alpha(x) = P_x e_\alpha$, $\alpha = 1, \ldots, d$. Then V_α's are smooth vector fields. Let $\varepsilon = 1$. In this case the solution to S.D.E.(3.1) is a Brownian motion on M (Stroock's construction !).

Note that $X(1,\cdot;\theta)^*_x : TM^*_{X(1,x;\theta)} \to TM^*_x$ is naturally extensible to $\Gamma(X(1,\cdot;\theta)^*_x): \Lambda TM^*_{X(1,x;\theta)} \to \Lambda TM^*_x$. By Berezin formula (c.f. Fadeev[2]), we have

$$\det_{TM_x}(I_{TM_x} - X(1,\cdot;\theta)_{*x}) = \int \exp(\, -a^*a + (a^*, X(1,\cdot;\theta)_{*x}a)_x)da^*da$$

$$= \mathrm{Strace}(\Gamma(X(1,\cdot;\theta)^*_x) \quad \text{for}$$

$\delta(X(1,x),x)m(dx)\otimes\mu(d\theta)$-a.e.$(x,\theta)$. Here

$$\mathrm{Strace}(A) = \mathrm{trace}\, A|_{\sum_{p:\mathrm{even}} \overset{p}{\Lambda} TM_x^*} - \mathrm{trace}\, A|_{\sum_{p:\mathrm{odd}} \overset{p}{\Lambda} TM_x^*}$$

for $A: \Lambda TM_x^* \to \Lambda TM_x^*$ and a, a^* are Grassman variables on $T_x(M)$. Let $\mathcal{F}_x$ be a σ-field generated by $\{X(t,x;\cdot),\ t\in[0,1]\}$. Then we have

$$\begin{aligned} &\deg(\Phi|_{\tilde{M}}) \\ &= \int_M m(dx)\ E^{\mu}[\ \varphi_g(X(\cdot,x;\theta))\delta(X(1,x;\theta),x) \\ &\qquad\qquad \mathrm{Strace}\ E^{\mu}[\ \Gamma(X(1,\cdot;\theta)_x^*)|\mathcal{F}_x]\ ] \end{aligned}$$

Set $M(t,w) = E^{\mu}[\ \Gamma(X(t,\cdot;\theta)\ |\mathcal{F}_x\ ]|_{X(\cdot,x;\theta)=w(\cdot)}$, $(t,w) \in [0,1]\times C([0,1];M)$. Then $M(t,w) \in \mathrm{Hom}(\Lambda TM_{w(t)}^*;\Lambda TM_{w(0)}^*)$. Let P_x, $x \in M$, be the Wiener measure on $C([0,1];M)$ starting from x, and let $P_t\beta \in \Gamma(M;\Lambda TM^*)$ for each $\beta \in \Gamma(M;TM^*)$ given by

$P_t\beta(x) = E^{P_x}[\ M(t,x,w)(\beta(w(t)))\ ]$, $x \in M$. Then we can prove that

$$\frac{\partial}{\partial t} P_t\beta(x) = \frac{1}{2}\ \square\ P_t\beta(x)$$

$$P_t\beta\ |_{t=0} = \beta\ .$$

Here $\square$ is the Hodge-Kodaira Laplacian. Let $P(t,x,y) \in \mathrm{Hom}(\Lambda TM_y^*;\Lambda TM_x^*)$ be the kernel of P_t. Also, Let $p(t,x,y)$ be the fundamental solution of the heat equation, $\frac{\partial}{\partial t} u = \frac{1}{2}\Delta u$. Then we see the following.

(3.5) $\chi(M) = \int_M m(dx)\ \mathrm{Strace}\ P(1,x,x)$,

(3.6) $\int_M m(dx)\ p(1,x,x)\ E^{P_x}[\ \mathrm{Strace}\ M(t,w),$

$\{w(t);t\in[0,1]\}$ is conjugate to $g\ |\ w(1) = x\ ]$

$$= \begin{cases} 0 & \text{if } g \neq e \\ \chi(M) & \text{if } g = e. \end{cases}$$

4. Proof.

Throughout this section we assume the assumptions (A-1) - (A-4) in Section 3 .

(4.1) Lemma. Let $h \in B^*$ with $\|h\|_H = 1$ and $\psi \in C_0^\infty(\mathbf{R})$ satisfying $\int_{\mathbf{R}} \psi(t)e^{-t^2/2}dt = 0$, and let $g(z) = \psi({}_B\langle z,h\rangle_{B^*})$, $z \in B$. Then $\int_{M\times B} g\circ\Phi\ d(\Phi|_{\tilde{M}})^*\mu = 0$.

Proof. We will prove this lemma in some steps.

Step 1. Let $\tilde{\psi}(t) = -e^{t^2/2}\cdot\int_{-\infty}^{t}\psi(s)e^{-s^2/2}ds$. Then $\tilde{\psi} \in C_0^\infty(R)$ and $\psi(t) = t\tilde{\psi}(t) - \frac{d}{dt}\tilde{\psi}(t)$. Let $\tilde{k}(z) = \tilde{\psi}({}_B\langle z,h\rangle_{B^*})$ and $k(z) = \tilde{k}(z)h$, $z \in B$. Then it is easy to see that

(4.2) $\partial k(z) = \tilde{k}(z)_B\langle z,h\rangle_{B^*} - (D\tilde{k}(z),h)_H = g(z)$.

Let $A(x,z) = (P_H+DV(x,z))(I - P(x,z)) + S(x,z)$

$= I + DV(x,z) + R_1(x,z)$

Note that $DV(x,z) + R_1(x,z)$ is a Hilbert-Schmidt operator in $TM_x \oplus H$. Let $u(x,z) = \det_2(A(x,z))A(x,z)^{-1}(0,h)$. Since $A(x,z)u(x,z) = \det_2(A(x,z))(0,h)$, we see that $S(x,z)u(x,z) = 0$ and so $P(x,z)u(x,z) = 0$. Therefore $(P_H+DV(x,z))u(x,z) = \det_2(A(x,z))h$. So we have

$$D(\tilde{k}\circ\Phi)(x,z)(u(x,z)) = D\tilde{k}(\Phi(x,z))(P_H+DV(x,z))u(x,z)$$
$$= (h,D\tilde{k}(\Phi(x,z)))_H \det_2(A(x,z)).$$

Let $f(x,z) = \exp(\,-\,\partial V(x,z) - \frac{1}{2}\|V(x,z)\|_H^2 + \operatorname{trace} R_1(x,z)\,)$.

Then we have

(4.3) $(D\tilde{k}\circ\Phi,h)\; d(\Phi|_{\tilde{M}})^*\mu$

$= \varphi(x,z)(D(\tilde{k}\circ\Phi)(x,z),f(x,z)u(x,z))_{TM_x\oplus H}\;\tilde{\nu}(dx\otimes dz)$.

Step 2. We will show that

(4.4) $\partial iv(f(x,z)u(x,z)) = {}_B\langle\Phi(x,z),h\rangle_{B^*}\det_2(A(x,z))f(x,z)$

for $\tilde{\nu}$-a.e.(x,z).

Since both sides of (4.4) are elements in $\mathcal{E}^\infty_{1+}(M;R)$, it is sufficient to show (4.4) for $m\otimes\mu$-a.e.(x,z). Let $\{P_n\}_{n=1}^\infty$ is a family of orthogonal projections in H satisfying $P_n\uparrow$ Identity and Image(P_n) $\subset B^*$. Let $V_n(x,z) = P_nV(x,z)$ and let

$A_n(x,z) = (P_H+DV_n(x,z))(I - P(x,z)) + S(x,z)$,

$u_n(x,z) = \det_2(A_n(x,z))A_n(x,z)^{-1}(0,h)$,

$R_{1,n}(x,z) = (P_H+DV_n(x,z))(I-P(x,z)) + S(x,z) - I$,

and $f_n(x,z) = \exp(\,-\,\partial V_n(x,z) - \frac{1}{2}\|V_n(x,z)\| + \operatorname{trace} R_{1,n}(x,z)\,)$.

Then since V and F have sufficient regularity, we see that

$\partial iv(f_n u_n)(x,z) \to \partial iv(fu)(x,z)$ in measure with respect to $dm\otimes d\mu$. Note that $f_n(x,z)u_n(x,z)$

$= \exp(-{}_B\langle z,P_nV(x,z)\rangle_{B^*} - \frac{1}{2}\|P_nV(x,z)\|_H^2)\ \det(A_n(x,z))A_n(x,z)^{-1}(0,h).$

Suppose that N be a finite dimensional Riemannian manifold and let $A \in \Gamma(N;TN\otimes TN)$ and $u \in \Gamma(N;TN)$. Then by easy calculation, we see that $\mathrm{div}(\det(A)A^{-1}u)(y) = \sum_{i=1}^{\dim N} (e_i, A^{-1}\nabla_{e_i}u)$, where $\{e_i\}$ is orthonomal basis of TN_y and ∇ is the contravariant derivative.

Therefore by noting that

$(P_H+P_nDV(x,z))(\det(A_n(x,z))A_n(x,z)^{-1}(0,h)) = \det(A_n(x,z))h,$

we see that

$$\begin{aligned}&\partial iv(\det(A_n(x,z))A_n(x,z)^{-1}(0,h))\\ &= {}_B\langle z,h\rangle_{B^*}\det(A_n(x,z))\\ &\qquad + {}_B\langle z,P_nDV(x,z)(\det(A_n(x,z))A_n(x,z)^{-1}(0,h))\rangle_{B^*}.\end{aligned}$$

On the other hand, we see that

$$\begin{aligned}&D(\ \exp(\ -\ {}_B\langle z,P_nV(x,z)\rangle_{B^*} - \tfrac{1}{2}\|P_nV(x,z)\|_H^2\ )\ )\\ &= \{\ -\ P_nV(x,z) - {}_B\langle z,P_nDV(x,z)\rangle_{B^*} - (P_nV(x,z),P_nDV(x,z))_H\ \}\\ &\qquad \times \exp(\ -\ {}_B\langle z,P_nV(x,z)\rangle_{B^*} - \tfrac{1}{2}\|P_nV(x,z)\|_H^2 - \mathrm{trace}\ P_nDV(x,z)\ ).\end{aligned}$$

Therefore we have

$$\begin{aligned}&\partial iv(f_nu_n)\\ &= \{(P_nV(x,z),(P_H+P_nDV(x,z))(\det(A(x,z))A(x,z)^{-1}(0,h)))_H\\ &\qquad\qquad + {}_B\langle z,h\rangle_{B^*}\det(A_n(x,z))\}\\ &\qquad \times \exp(\ -\ {}_B\langle z,P_nV(x,z)\rangle_{B^*} - \tfrac{1}{2}\|P_nV(x,z)\|_H^2\ )\\ &= {}_B\langle z+P_nV(x,z),h\rangle_{B^*}\det{}_2(A_n(x,z))f_n(x,z)\\ &\to {}_B\langle \Phi(x,z),h\rangle_{B^*}\det{}_2(A(x,z))f(x,z)\end{aligned}$$

in measure with respect to $dm\otimes d\mu$. This proves (4.4).

Step 3. Let $\{g_n\}_1^\infty$ be as in the assumption (A-4). By the assumption (A-3), we see that $g_n(x)\varphi(x,z)f(x,z)u(x,z) \in \mathcal{E}_{1+}^\infty(M;H)$. Since $S(x,z)u(x,z) = 0$, we have $E(x,z)u(x,z) = u(x,z)$. Also we have $(D\varphi(x,z),u(x,z))_{\mathcal{H}(H)} = 0$. Therefore by Proposition(2.1) and

Lemma(2.3), we have

$$\int_{M\times B} g_n(x)\partial k(\Phi(x,z))d(\Phi|_{\tilde{M}})^*\mu$$

$$= -\int_{M\times B} g_n(x)(D(\tilde{k}\circ\Phi)(x,z),\varphi(x,z)f(x,z)u(x,z))_{TM_x\oplus H}\,\tilde{\nu}(dx\otimes dz)$$

$$+\int_{M\times B} g_n(x)\tilde{k}(\Phi(x,z)_B\langle\Phi(x,z),h\rangle_{B^*}\det{}_2(A(x,z))\varphi(x,z)f(x,z)\tilde{\nu}(dx\otimes dz)$$

$$= -\int_{M\times B}(\tilde{k}\circ\Phi)(x,z)\partial iv(g_n(x)\varphi(x,z)f(x,z)u(x,z))\,\tilde{\nu}(dx\otimes dz)$$

$$+\int_{M\times B} g_n(x)(\tilde{k}\circ\Phi)(x,z)_B\langle\Phi(x,z),h\rangle_{B^*}\det{}_2(A(x,z))\varphi(x,z)f(x,z)\tilde{\nu}(dx\otimes dz)$$

$$= \int_{M\times B}(\mathrm{grad}\ g_n(x),\mathrm{Proj}_{TM_x}u(x,z))_{TM_x}(\tilde{k}\circ\Phi)(x,z)f(x,z)\tilde{\nu}(dx\otimes dz).$$

Thus we see that

$$\left|\int_{M\times B} g_n(x)\partial g(\Phi(x,z))d(\Phi|_{\tilde{M}})^*\mu\ \right|$$

$$\leqq \|\tilde{\psi}\|_\infty\cdot\int_{M\times B}\|\mathrm{grad}\ g_n(x)\|_x\|A(x,z)^{-1}\|_{\mathrm{operator}}|d(\Phi|_{\tilde{M}})^*\mu| \to 0\ ,$$

as $n\to\infty$. This proves our assertion.

(4.5)<u>Corollary</u>. $\int_{M\times B} f\circ\Phi\ d(\Phi|_{\tilde{M}})^*\mu = \deg(\Phi|_{\tilde{M}})\int_B f\ d\mu$

for any bounded measurable function f on B.

<u>Proof</u>. Take $\{\psi_n\}_{n=1}^{\infty}\subset C_0^\infty(\mathbf{R})$ satisfying $\psi_n \geqq 0$ and $\psi_n(t)\uparrow 1$ for each $t\in\mathbf{R}$. For each $h\in B^*$ with $h\neq 0$, let $g_h(z) = \exp(\sqrt{-1}\,_B\langle z,h\rangle_{B^*})$, $z\in B$. Then by Lemma(4.1), we have

$$\int_{M\times B} g_h\circ\Phi\ d(\Phi|_{\tilde{M}})^*\mu$$

$$= \lim_{n\to\infty}\int_{M\times B}((g_h\psi_n(_B\langle\cdot,h\rangle_{B^*})\circ\Phi - (\int_B g_h(z)\psi_n(_B\langle z,h\rangle_{B^*})\mu(dz)))\ d(\Phi|_{\tilde{M}})^*\mu$$

$$+ \lim_{n\to\infty}\int_{M\times B}(\int_B g_h(z)\psi_n(_B\langle z,h\rangle_{B^*})\mu(dz))\ d(\Phi|_{\tilde{M}})^*\mu$$

$$= \deg(\Phi|_{\tilde{M}})\int_B g_h\ d\mu.$$

Therefore (4.5) holds for $f = \exp(\sqrt{-1}\,_B\langle\cdot,h\rangle_{B^*})$, $h\in B^*$. Thus by the routin argument we see that (4.5) holds for all bounded measurable function f. Q.E.D.

(4.6)<u>Lemma</u>. There is a measurable set A in B satisfies the following.

(1) $\mu(A) = 1$, and $A + H = A$.

(2) There is a family of countably many compact sets $\{K_n\}_{n=1}^{\infty}$ in $M\times B$

such that

$$\bigcup_{n=1}^{\infty} K_n = \{(x,z) \in M\times A;\ F(x,z) = x,\ \det(DF(x,z)DF(x,z)^*) \neq 0,$$
$$\det{}_2((P_H+DV(x,z))(I-P(x,z))+S(x,z)) \neq 0$$
$$\text{and } \varphi(x,z) > \tfrac{1}{2}\},$$

$\Phi|_{K_n}: K_n \to B$ is one to one, and $|(\Phi|_{\tilde{M}})^*\mu| = \mu\circ(\Phi|_{K_n})$ on K_n.

(4.7)<u>Lemma</u>. $\Phi(\{(x,z)\in M\times B;\ P_H+DV(x,z):TM_x\oplus H\to H$ is not a onto map $\}) \subset B$ is of μ-measure zero.

Lemmas (4.6) and (4.7) are proved similarly to [6] (see also [7]).

(4.8)<u>Lemma</u>. $\Phi(\{(x,z)\in M\times B;\ \det(DF(x,z)DF(x,z)^*) = 0\}) \subset B$ is of μ-measure zero.

<u>Proof</u>. From Lemma(4.7), it is sufficient to show that $\Phi(\{(x,z)\in M\times B;$ $\det(DF(x,z)DF(x,z)^*)=0$ and $P_H+DV(x,z):TM_x\oplus H\to H$ is non-degenerate$\})$ is of μ-measure zero. But similarly to [6], we can find a measurable set A in B such that

(1) $\mu(A) = 1$ and $A + H = A$, and

(2) there is a family of countably compact $\mathcal{H}$-C^0 maps $\{\varphi_n\}_{n=1}^{\infty}$ from $M\times B$ into $\mathbb{R}$ such that

$$\bigcup_{n=1}^{\infty} \{(x,z)\in M\times B;\ \varphi_n(x,z) > 0\}$$

$= \{(x,z) \in M\times A;\ I_{M\times B}+DV(x,z):TM_x\otimes H\to TM_x\otimes H$ is one to one onto$\}$ and

and $(I_{M\times B}+V)|_{\{\varphi_n>0\}}:\{(x,z)\in M\times B;\ \varphi_n(x,z)>0\} \to M\times B$ is one to one.

Since $\mathrm{Cap}_{1,p}(\{(x,z)\in M\times B;\det(DF(x,z)DF(x,z)^*)=0\}) = 0$ for any $p \in (1,\infty)$, we see that $A_n = (I_{M\times B}+V)(\{(x,z)\in M\times B;\varphi_n(x,z)>0$ and $\det(DF(x,z)DF(x,z)^*)=0\})$ is 'slim' in $M\times B$. Therefore projection to B of $\bigcup_n A_n$ is of μ-measure zero. This proves our assertion. See [7] for the detail. Q.E.D.

Now let us prove Theorem(2.6). Let $\{K_n\}_{n=1}^{\infty}$ be as in Lemma(4.7).

Let $E_1 = K_1$ and $E_{n+1} = K_{n+1} \setminus \bigcup_{k=1}^{n} K_k$, $n \geqq 1$. Then by Corollary(4.5), we have for any bounded measurable function f in B

$$\deg(\Phi|_{\tilde{M}}) \int_B f \, d\mu$$
$$= \int_{M\times B} f\circ\Phi \, d(\Phi|_{\tilde{M}})^* \mu$$
$$= \sum_{n=1}^{\infty} \int_{E_n} f\circ\Phi(x,z) \, \mathrm{sign}(\det_2(A(x,z))) \, \mu\circ(\Phi|_{E_n})(dx\otimes dz)$$
$$= \int_B f(w) \sum_{F(x,z)=w} \chi_{\underset{n}{\cup} E_n}(x,z) \, \mathrm{sign}(\det_2(A(x,z))) \, \mu(dw),$$

where $A(x,z) = (P_H+DV(x,z))(I-P(x,z))+S(x,z)$.

Therefore by this and Lemmas (4.7) and (4.8), we see that

$$\deg(\Phi|_{\tilde{M}}) = \sum_{\Phi(x,z)=w} \chi_{\underset{n}{\cup} E_n}(x,z) \, \mathrm{sign}(\det_2(A(x,z)))$$
$$= \sum_{\substack{\Phi(x,z)=w \\ F(x,z)=x \\ \varphi(x,z)=1}} \mathrm{sign}(\det_2(A(x,z)))$$

for μ-a.e.w. This proves our Theorem.

References.

[1] Bismut, J-M., Index theorem and equivariant cohomology on the loop space, Comm. Math. Phys. 98(1985), 213-237.

[2] Fadeev, L., Introduction to functional methods, Methods in field theory, Les Houches Session XXVIII 1975, pp. 1-40, North-Holland 1976.

[3] Getzler, E., Degree theory for Wiener maps, J. Func. Anal. 68(1986), 388-403.

[4] Ikeda, N., and S. Watanabe, Malliavin calculus of Wiener functionals and its applications, From local times to global geometry, control and physics, ed. K.D.Elworthy, Pitman Research Notes in Math. Series 150, Longnan Scientific & Technical 1976, Essex.

[5] Kree, P., Théores des distributions et calculs differentiels sur un espace des Banach, Seminarie P. Lelong (Analyse) 1974/75 pp. 163-192 Lect. Notes in Math. vol. 524, Springer 1976, Berlin.

[6] Kusuoka, S., Some remarks on Getzler's degree theorem, to appear

in Proc. of 5th Japan-USSR symp.

[7] Kusuoka, S., On Wiener-Riemannian manifolds, to be submitted to Proc. of Warwick Open House on Stochastic Analysis 1987.

[8] Malliavin, P., Implicit functions in finite corank on the Wiener space, Stochastic Analysis Proc. of Taniguchi Intn. Symp. on Stoch. Anal. (Katata and Kyoto 1982) ed. K. Ito, pp. 369-386, Kinokuniya Tokyo, 1984. (See also the article in this Proceedings.)

[9] Sugita, H., Positive generalized Wiener functions and potential thory over abstract Wiener spaces, to appear.

[10] Takanobu, S., Diagonal short time asymptotics of heat kernals for certain degenerate second order differential operators of Hörmander type, to appear in Publ. RIMS Kyoto Univ.

[11] Watanabe, S., Stochastic differential equations and Malliavin calculus, Tata Lecture notes, Springer Verlag 1984, Berlin, Heidelberg, New York, Tokyo.

APPLICATIONS QUANTITATIVES ET GEOMETRIQUES DU CALCUL DE MALLIAVIN

Rémi LEANDRE
Département de Mathématiques. Faculté des Sciences de Besançon
25030 Besançon Cédex (France)

I - Notions heuristiques de submersion au sens faible et au sens fort :

I.1) Le cas de la dimension finie :

Considérons une application F C^∞ de $\mathbb{R}^n$ dans $\mathbb{R}^d$. L'élément générique de $\mathbb{R}^n$ est noté w, celui de $\mathbb{R}^d$ y, l'application linéaire en w de F est notée DF(w). Rappelons que F est une submersion en w si DF(w) est une surjection de $\mathbb{R}^n$ dans $\mathbb{R}^d$. Ceci se traduit par le fait que la matrice de Gram $DF(w)^t\, DF(w)$, qui est une matrice symétrique positive sur $\mathbb{R}^d$, est définie. Nous dirons que F est une submersion au sens fort (en w_o sur $\mathbb{R}^n$) si la matrice de Gram de F est inversible (en w_o ; en tout point de $\mathbb{R}^n$).

Munissons maintenant $\mathbb{R}^n$ de sa structure d'espace euclidien canonique, la norme étant notée $\|\ \|$, et considérons la mesure gaussienne dP(w) non dégénérée sur $\mathbb{R}^n$ de densité $\frac{1}{(\sqrt{2\pi})^n} \exp\left[-\frac{\|w\|^2}{2}\right]$ par rapport à la mesure de Lebesgue sur $\mathbb{R}^n$. Nous dirons que F est une submersion au sens faible sur $\mathbb{R}^n$ si pour tout entier $p > 0$,

$$(1.1) \qquad E[(\det(DF(w)^t\, DF(w)))^{-p}] < \infty$$

et nous dirons que F est une submersion au sens faible au point w_o si il existe une fonction continue $g \geq 0$, strictement positive en w_o, de $\mathbb{R}^n$ dans $\mathbb{R}^+$, telle que pour tout entier $p > 0$

$$(1.1)' \qquad E[g(w)(\det\,(DF(w)^t\, DF(w)))^{-p}] < \infty.$$

Le fait remarquable en dimension finie est le suivant : les deux notions précédentes du mot submersion sur $\mathbb{R}^n$ sont équivalentes, à des conditions techniques près sur le comportement à l'infini de F. En tous cas, les deux notions précédentes du mot submersion en w_o le sont.

Par la suite, nous n'insisterons pas sur les difficultés techniques qui surgissent du comportement à l'infini de F, puisqu'il s'agit d'un exposé de synthèse et d'exposition.

Soit G_o une fonction C^∞ de $\mathbb{R}^n$ dans $\mathbb{R}$: nous éviterons encore de préciser son comportement à l'infini, pour les mêmes raisons. Soit f une fonction C^∞ de $\mathbb{R}^d$ dans $\mathbb{R}$, à support compact. Considérons la mesure :

$$(1.2) \qquad f \to E[G_o(w)\ f(F(w))].$$

Puisque F est une submersion (au sens faible ou au sens fort) sur $\mathbb{R}^n$, cette mesure possède une densité C^∞ $p(y)\,dy$ par rapport à la mesure de Lebesgue sur $\mathbb{R}^d$. On peut le voir de deux façons :

- On utilise le fait que l'on a une submersion au sens fort. L'image réciproque $F^{-1}(y)$ de y est une sous-variété de $\mathbb{R}^n$, et lorsque y décrit l'espace $\mathbb{R}^d$, on obtient un feuilletage de $\mathbb{R}^n$. On désintègre la mesure gaussienne sur $\mathbb{R}^n$. On obtient ([B.1] formule 0.7)

$$(1.3) \qquad p(y) = \frac{1}{(\sqrt{2\pi})^n} \int_{F^{-1}(y)} G_o(w) \frac{\exp\left[-\frac{\|w\|^2}{2}\right]}{(\det DF(w)^t\ DF(w))^{\frac{1}{2}}}\, d\sigma^y(w),$$

$d\sigma^y$ désignant la mesure de Lebesgue sur $F^{-1}(y)$.

- On utilise le fait que l'on a une submersion au sens faible. Cela nous permet d'obtenir des formules d'intégration par parties. Plus précisément, soit G une fonction C^∞ de $\mathbb{R}^n$ dans $\mathbb{R}$, dont toutes les dérivées sont dans tous les $L^p(dP)$. Soit (α) un multi-indice sur $\mathbb{R}^d$. Il existe une fonctionnelle $G(\alpha)$ telle que pour toute fonction C^∞ de $\mathbb{R}^d$ dans $\mathbb{R}$ à support compact :

$$(1.4) \qquad E[f^{(\alpha)}(F(w))\ G(w)] = E[f(F(w))\ G(\alpha)(w)].$$

Ceci nous montre que toutes les dérivées au sens des distributions de la mesure μ sont des mesures, et donc que μ est représentée par une densité C^∞.

Le problème qui nous intéresse par la suite est de trouver des estimations quand $\varepsilon \to 0$ de la densité $p_\varepsilon(y)$ de la mesure

$$(1.4) \qquad \mu_\varepsilon : f \to E[f(F(\varepsilon w))\ G(\varepsilon,w)],$$

$G(\varepsilon,w)$ étant une fonction C^∞ de $[0,1] \times \mathbb{R}^n$ uniformément intégrable en ε dans tous les $L^p[dP]$, et F étant une submersion au sens fort (ou faible, peu importe dans notre situation).

On peut utiliser à cette fin les deux points de vue précédents :

- on utilise le fait que F est une submersion <u>au sens fort</u>. On obtient une expression explicite de $p_\varepsilon(y)$:

$$(1.5)\qquad p_\varepsilon(y) = \frac{1}{(\sqrt{2\pi\varepsilon})^n} \int_{F^{-1}(y)} G(\varepsilon,w) \frac{\exp[-\frac{\|w\|^2}{2\varepsilon^2}]}{(\det\ DF(w)^t DF(w))^{\frac{1}{2}}} d\sigma^y(w)$$

et on effectue un développement sous le signe $\int$.

- on utilise le fait que F est une submersion au sens faible, et on étudie d'abord le cas où $0\in F^{-1}(y)$. Soit f_m une suite de fonctions de $\mathbb{R}^d$ dans $\mathbb{R}$ tendant au sens des distributions vers la masse de Dirac en 0. On a :

$$(1.6)\qquad \begin{aligned} p_\varepsilon(y) &= \lim_{m\to+\infty} E[f_m(F(\varepsilon w) - F(0))\ G(\varepsilon,w)] = \\ &= \frac{1}{\varepsilon^d} \lim_{m\to+\infty} E[f_m(\frac{F(\varepsilon w) - F(0)}{\varepsilon})\ G(\varepsilon,w)] = \varepsilon^{-d}\ \tilde{p}_\varepsilon(0), \end{aligned}$$

$\tilde{p}_\varepsilon(y)$ désignant la densité de la mesure $f \xrightarrow{\tilde{\mu}_\varepsilon} E[f(\frac{F(\varepsilon w) - F(0)}{\varepsilon})\ G(\varepsilon,w)]$. De plus quand $\varepsilon \to 0$, $\frac{F(\varepsilon w) - F(0)}{\varepsilon} \to DF(0).w$ qui est une variable gaussienne non dégénérée sur $\mathbb{R}^d$, car DF(0) est une surjection de $\mathbb{R}^n$ dans $\mathbb{R}^d$, car toute submersion au sens faible est encore une submersion au sens fort. De plus, on a un développement asymptotique par la formule de Taylor : pour tout entier N, il existe des fonctionnelles $L_j^{(\alpha)}$ $j\leq N$, telles que pour toute fonction f C^∞ de $\mathbb{R}^d$ dans $\mathbb{R}$ à support compact, on ait :

$$(1.7)\qquad \begin{aligned} f(\frac{F(\varepsilon w) - F(0)}{\varepsilon})\ G(\varepsilon,w) &= \sum_{j\leq N} \varepsilon^j \\ &\sum_{|\alpha|\leq j} f^{(\alpha)}\ (DF(0).w)\ L_j^{(\alpha)}(w) + \varepsilon^N \text{ reste.} \end{aligned}$$

En utilisant des formules du type de (1.4) et le fait que DF(0) est surjective, on peut faire disparaître les dérivées de f qui apparaissent dans (1.7) et dans le reste multiplié par ε^N. On obtient ainsi un développement asymptotique de la fonction $\varepsilon \to E[f(\frac{F(\varepsilon w) - F(0)}{\varepsilon})\ G(\varepsilon,w)]$ dans lequel ne figure aucune dérivée de f. On en déduit alors un développement asymptotique en 0 de la densité $\tilde{p}_\varepsilon(.)$ de la mesure $\tilde{\mu}_\varepsilon$, et par suite un développement asymptotique de $\tilde{p}_\varepsilon(0)$. (1.6) nous permet alors de conclure.

Quand $F^{-1}(y)$ ne contient pas 0, on effectue le changement de variable $w \to w + \frac{w_o}{\varepsilon}$, w_o étant un élément de $F^{-1}(y)$ de norme minimum. Par suite, $F(\varepsilon w)$ est transformé en $F(\varepsilon w + w_o)$ et la mesure $dP(w)$ en $\exp[-\frac{\|w_o\|^2}{2\varepsilon^2}]\ \exp[-\frac{\langle w_o,w\rangle}{\varepsilon}]\ dP(w)$. On obtient alors :

$$(1.8) \qquad p_\varepsilon(y) = \exp\left[-\frac{\|w_o\|^2}{2\varepsilon^2}\right] \bar{p}_\varepsilon(0),$$

$\bar{p}_\varepsilon(.)$ étant la densité de la mesure :

$$(1.9) \qquad \begin{aligned} & f \xrightarrow{\bar{\mu}_\varepsilon} E[f(F(\varepsilon w + w_o) - F(w_o))\, \bar{G}(\varepsilon,w)] \\ & \bar{G}(\varepsilon,w) = G(\varepsilon,w) \exp\left[-\frac{\langle w_o,w\rangle}{\varepsilon}\right]. \end{aligned}$$

On s'est ramené au cas précédent. Toutefois, il subsiste un problème majeur : il faut contrôler le terme exponentiel dans $\bar{G}(\varepsilon,w)$. On utilise à cette fin le fait qu'une submersion au sens faible en w_o est une submersion au sens fort en w_o. $F^{-1}(y)$ est donc une sous-variété au voisinage de w_o. Pour minimiser la norme sur ce voisinage, on peut appliquer la méthode des multiplicateurs de Lagrange. Cela prouve que $\langle w_o,w\rangle = 0$ dès que $DF(w_o)(w) = 0$.

I.2) <u>Une rupture en dimension infinie</u> :

Le seul problème en dimension finie consiste à contrôler au voisinage de l'infini le comportement des dérivées des fonctionnelles que l'on considère. En dimension infinie, la situation devient beaucoup plus complexe, comme le suggère l'exemple suivant : considérons un χ^2 à n degrés de libertés. $F = \sum_{i=1}^{n} w_i^2$ $w = (w_1,\ldots,w_n)$, $\mathbb{R}^d = \mathbb{R}$. En 0, ce n'est <u>jamais</u> une submersion. Toutefois, il est bien connu qu'il existe une suite $p_n \to \infty$ lorsque $n \to \infty$ tel que :

$$(1.10) \qquad E[(DF(w)\ {}^tDF(w))^{-p_n}] < \infty.$$

Ceci corrobore d'ailleurs le fait que la densité d'un χ^2 à n degrés de libertés devient de plus en plus régulière lorsque le nombre de degrés de libertés croît.

L'exemple suivant est dû à J.M. Bismut ([B.1]). Soit H_2 l'espace de Cameron-Martin sur $\mathbb{R}^m$, c'est-à-dire l'ensemble des fonctions $t \to (h_{i,t})$ de $[0,1]$ dans $\mathbb{R}^m$ de carré intégrable. H_2 est muni de la norme énergie

$$(1.11) \qquad \|h\|^2 = \sum_{i=1}^{m} \int_0^1 h_{i,t}^2\, dt.$$

Introduisons m champs de vecteurs $X_1,\ldots,X_m$ sur $\mathbb{R}^d$ de dérivées de tout ordre bornées, et considérons la solution de l'équation différentielle

$$(1.12) \qquad \begin{aligned} & dx_t(h) = \sum_{i=1}^{m} X_i(x_t(h))\, h_{i,t}\, dt \\ & x_o(h) = x. \end{aligned}$$

Posons $\Phi(h) = x_1(h)$. On obtient ainsi une application C^∞ de H_2 dans $\mathbb{R}^d$, et on vérifie que Φ est une submersion au sens fort sur H_2 (en 0) si et seulement si l'espace engendré par les champs $X_i(x)$ est égal à $\mathbb{R}^d$. Toutefois, on ne peut donner immédiatement un sens à la notion de submersion au sens faible pour Φ, car il est impossible de munir H_2 de la mesure gaussienne $\exp\left[-\frac{\|h\|^2}{2}\right]$ "dh". C'est pourquoi nous devons changer de fonctionnelle. Soit un mouvement brownien m-dimensionnel $(w_1,\ldots,w_m)$ et soit l'équation différentielle de Stratonovitch sur $\mathbb{R}^d$:

$$(1.13) \qquad \begin{aligned} dx_t(dw) &= \Sigma\, X_i(x_t(dw))\, dw_i \\ x_o(dw) &= x. \end{aligned}$$

En posant $F(w) = x_1(dw)$, nous obtenons une fonctionnelle brownienne qui n'est pas C^∞ au sens usuel de la topologie de la norme uniforme sur les trajectoires browniennes, mais qui est C^∞ au sens de Malliavin (nous renvoyons à [M] et aux références contenues dans [M] sur ce sujet). La matrice de Graham $DF(w)\ {}^tDF(w)$ est appelée dans ce contexte matrice de Malliavin, et on obtient une submersion au sens faible si et seulement si pour tout entier $p > 0$

$$(1.14) \qquad E[(DF(w)\ {}^tDF(w))^{-p}] < \infty.$$

En particulier (1.14) est réalisée dès que l'algèbre de Lie engendrée en x par tous les champs X_i est égale à $\mathbb{R}^d$ ([M], [H] : hypothèse forte de Hörmander).

Dans ce cas, le problème de l'estimation quand $\varepsilon \to 0$ de la densité $p_\varepsilon(x,y)$ de $F(\varepsilon w)$ va donner lieu à deux types de méthodes différentes, F et Φ intervenant à des niveaux différents.

La première est due essentiellement à J.M. Bismut ([B.1]) : elle utilise la notion de submersion au sens fort sur Φ.

La deuxième utilise la notion (heuristique dans notre contexte) de submersion au sens faible sur F : elle fait intervenir Φ et F à des niveaux différents. Ainsi, presque toutes les trajectoires browniennes sont d'énergie infinie : w_o dans l'analogue de (1.8) sera donc calculé pour Φ dans $\Phi^{-1}(y)$, et non pour F dans $F^{-1}(y)$. (1.8) sera la conséquence d'une formule de Girsanov, classiquement utilisée dans ce genre de problèmes ([Mol], [K], [Az.1]) et $\langle w_o, w\rangle$ deviendra une intégrale stochastique. L'analogue de la formule (1.7) a été obtenu auparavant par R. Azencott ([Az.2]), H. Doss ([D]) alors que l'idée de faire disparaître les dérivées de f dans l'expression (1.7) intégrée au moyen du calcul de Malliavin est due indépendamment à S. Watanabe ([W]) et à nous-mêmes ([L.1], [L.2]). Toutefois, diviser par ε dans (1.6), comme il était d'usage jusqu'à présent ([Mol], [K], [Az.1] et [Gr] en analyse) ne suffira sans doute pas, car DF(0)(w) n'a aucune raison d'être une gaussienne non dégénérée. De même, la méthode des multiplicateurs de Lagrange sera inapplicable. Il est à noter que G. Ben Arous a resitué cette méthode dans l'esprit de la méthode de la phase stationnaire ([B.A.1]).

II - Utilisation de la notion heuristique de submersion au sens faible dans l'estimation de la densité d'une diffusion hypoelliptique :

II.1) Les estimations de Varadhan :

Considérons l'opérateur sur $\mathbb{R}^d$:

$$L = \sum_{i=1}^{m} X_i^2 + X_o. \tag{2.1}$$

Supposons que les champs de vecteurs X_i $i = 0,\ldots,m$ sur $\mathbb{R}^d$ possèdent des dérivées de tout ordre. Supposons de plus qu'en tout point x, l'algèbre de Lie engendrée par les champs X_i $i \neq 0$ est égale à $\mathbb{R}^d$. Dans ce cas, le semi-groupe P_t associé à L possède une densité C^∞ $p_t'(x,y)$ pour $t > 0$ ([H]). On représente stochastiquement $p_t(x,y)$ en posant $\varepsilon^2 = t$, et on introduit la solution de l'équation différentielle stochastique de Stratonovitch

$$(2.2)\qquad \begin{aligned} dx_s(\varepsilon,dw) &= \varepsilon \sum_{i=1}^{m} X_i(x_s(\varepsilon,dw))\, dw_i + \varepsilon^2 X_o(x_s(\varepsilon,dw))\, ds \\ x_o(\varepsilon,dw) &= x. \end{aligned}$$

On a alors $p_t'(x,y) = p_\varepsilon(x,y)$, $p_\varepsilon(x,y)$ désignant la densité de la loi de $x_1(\varepsilon,dw)$. La solution d'une équation du type (1.12) est appelée une courbe horizontale. Il résulte de notre hypothèse sur les crochets de Lie que l'on peut rejoindre x à y par une courbe horizontale ; en d'autres termes $\Phi^{-1}(y) \neq 0$ ([B.1], Chap. 1). L'analogue dans (1.8) de $\inf_{\Phi^{-1}(y)} \|h\|^2$ est noté classiquement $d^2(x,y)$ ([B.1], [S], [Az.3]). De plus le minimum de l'énergie sur $\Phi^{-1}(y)$ est atteint au moins en un élément h_o de H_2 ([Az.3]), correspondant dans notre contexte à w_o dans (1.8). Malheureusement, Φ n'a aucune raison d'être une submersion en h_o. Cependant, on peut contourner ce problème en considérant un point suffisamment proche de h_o dans $\Phi^{-1}(y)$ dans H^2 où Φ est une submersion. Ceci nous permet d'obtenir le théorème suivant :

Théorème II.1 (Estimation de Varadhan) ([L.3], [L.4], [L.5]) : Uniformément sur tout compact, on a :

$$\lim_{\varepsilon\to 0} 2\varepsilon^2 \operatorname{Log} p_\varepsilon(x,y) = -d^2(x,y). \tag{2.3}$$

Preuve : Nous donnons le schéma de la preuve de l'inégalité

$$\underline{\lim}_{\varepsilon\to 0} 2\varepsilon^2 \operatorname{Log} p_\varepsilon(x,y) \geq -d^2(x,y). \tag{2.4}$$

La remarque essentielle est qu'il existe pour $\eta > 0$ un élément $h'(\eta)$ de $\Phi^{-1}(y)$ tel que :

* $\|h'(\eta)\|^2 \leq d^2(x,y) + \eta$

** Φ est une submersion au sens fort en $h'(\eta)$.

On effectue dans (2.2) la translation $dw \to dw + h_t'(\eta)dt/\varepsilon$, $x_1(\varepsilon,dw)$ est ainsi transformé en $x_1(\varepsilon,w,h)$, et on introduit une fonction troncatrice χ C^∞ de $\mathbb{R}$ dans $[0,1]$ égale à 0 en dehors de $[-\eta,\eta]$ et égale à 1 en 0. δw_i désignant la différentielle d'Ito, introduisons la mesure $\bar{\mu}_\varepsilon$ sur $\mathbb{R}^d$:

$$(2.5) \qquad f \to E[\chi(\varepsilon \int_0^1 \sum_{i=1}^m h'_{i,s}(\eta)\, \delta w_i)\, f(x_1(\varepsilon,w,h) - y)].$$

(1.8), du fait de l'adjonction dans $\bar{\mu}_\varepsilon$ de la fonction χ est transformée en

$$(2.6) \qquad p_\varepsilon(x,y) \geq \exp\left[-\frac{d^2(x,y) + 2\eta}{2\varepsilon^2}\right] \bar{p}_\varepsilon(0),$$

$\bar{p}_\varepsilon(z)$ étant la densité de $\bar{\mu}_\varepsilon$. On applique sur $\bar{\mu}_\varepsilon$ la procédure de décomposition de l'espace de Wiener de J.M. Bismut ([B.1], [L.5]). On décompose w_t en $w_t^1 + w_t^2$. w_t^2 est "formellement" la projection de w_t sur $(\operatorname{Ker} D\Phi(h'(\eta)))^\perp$ et w_t^1 celle de $w_.$ sur $(\operatorname{Ker} D\Phi(h'(\eta)))$. On a ainsi :

$$(2.7) \qquad w_{2.} = {}^tD\Phi(h'(\eta))(D\Phi(h'(\eta))^t\, D\Phi(h'(\eta)))^{-1}\, D\Phi(h'(\eta))\, dw.$$

$w_.^1$ et $w_.^2$ sont deux processus gaussiens indépendants, la loi de $w_.^2$ étant en fait une mesure gaussienne sur un sous-espace de H_2 de dimension finie, et celle de $w_.^1$, notée $dP_1(dw^1)$ étant la mesure cylindrique gaussienne associée au sous-espace de Hilbert de H_2 tangent à $\Phi^{-1}(y)$ en $h'(\eta)$. Puisque Φ est une submersion en $h'(\eta)$, on peut paramétrer un petit voisinage de $h'(\eta)$ dans $\Phi^{-1}(y)$ par un petit voisinage de 0 dans H_2 de façon C^∞, car $\Phi^{-1}(y)$ est au voisinage de $h'(\eta)$ une sous-variété de H_2. Mais comme nous considérons F, on ne peut le faire qu'approximativement pour $F^{-1}(y)$. L'approximation obtenue de $\bar{p}_\varepsilon(0)$ est de la forme $\int (\ldots)\, dP_1(dw^1)$ (nous omettons d'écrire la quantité $(\ldots)$, dw^1 mesurable), qui est à rapprocher de (1.3). Il ne reste plus qu'à montrer que cette approximation minore $\bar{p}_\varepsilon(0)$, est positive > 0, et que sa limite inférieure quand $\varepsilon \to 0$ est non nulle. ■

Remarque I : Dans [L.4], on utilise la notion heuristique de submersion au sens faible pour obtenir (2.4). Dans [L-R], on démontre des estimations de Varadhan pour la densité de diffusions indexée par un temps à plusieurs paramètres, en utilisant la notion de submersion au sens faible. On pourrait le faire aussi pour la notion de submersion au sens fort.

Remarque II : On peut rapporcher ces résultats de ceux obtenus par des méthodes différentes lorsque X_0 appartient à l'espace engendré par les crochets de Lie d'ordre ≤ 2 construits à partir des X_i, $i \neq 0$ ([K-S.2], [F-S], [J-S], [N-S-W], [S], [V]). Contrairement aux nôtres, ils sont valides pour tout $\varepsilon > 0$, et sont de la forme :

$$\frac{C'}{\varepsilon^{N'}} \exp\left[-\frac{d^2(x,y)}{C_1' \varepsilon^2}\right] \leq p_\varepsilon(x,y) \leq \frac{C}{\varepsilon^N} \exp\left[-\frac{d^2(x,y)}{C_1 \varepsilon^2}\right], \tag{2.8}$$

N étant le grade de l'algèbre de Lie engendrée par les X_i (cf. la IVème partie). Comme nous le notons ([L.4] et [L.5]), ces résultats, la théorie des grandes déviations ([Az.3]), et la formule de Kolmogorov permettent de déduire (2.3). Toutefois, (2.8) n'est plus vraie quand X_o n'est pas dans l'espace engendrée par les crochets de Lie d'ordre ≤ 2 construits à partir des X_i, comme le montre l'exemple suivant :

$$\mathbb{R}^d = \mathbb{R}^2, \quad X_o = \begin{pmatrix} 0 \\ 1 \end{pmatrix}, \quad X_1 = \begin{pmatrix} 1 \\ 0 \end{pmatrix}, \quad X_2 = \begin{pmatrix} 0 \\ x_1^3 \end{pmatrix}, \quad x = \begin{pmatrix} 0 \\ 0 \end{pmatrix},$$

$x_1(\varepsilon, dw) = \begin{pmatrix} w_1 \varepsilon \\ \varepsilon^2 + \varepsilon^4 \int_o^1 w_{1,s}^3 \, dw_2 \end{pmatrix}$. Dans ce cas $p_\varepsilon = \frac{1}{\varepsilon^5} \cdot \bar{p}\left(\begin{pmatrix} 0 \\ -\frac{1}{\varepsilon^2} \end{pmatrix}\right)$, $\bar{p}((.))$ étant la densité de la variable aléatoire sur $\mathbb{R}^2 \begin{pmatrix} w_1 \\ \int_o^1 w_{1,s}^3 \, dw_2 \end{pmatrix}$, qui est à décroissance rapide. Cet exemple montre que l'on peut conjecturer que, lorsque X_o n'est pas dans le sous-espace engendré par les crochets de Lie d'ordre 2, l'on a une décroissance rapide de $p_\varepsilon(x,x)$. Nous renvoyons à ce sujet à un article en préparation écrit en collaboration avec G. Ben Arous.

II.2) <u>Développement asymptotique de la densité en dehors du cut-locus</u> :

Nous dirons que nous sommes en dehors du cut-locus de L si le couple (x,y) de points de $\mathbb{R}^d$ vérifie les 3 conditions suivantes :

* Il existe un <u>unique</u> $h \in \Phi^{-1}(y)$ tel que $d^2(x,y) = \|h\|^2$

** $D\Phi(h)$ est une <u>surjection</u>, ce qui implique que $\Phi^{-1}(y)$ est au voisinage de h une sous-variété

*** $\|h\|^2$ est un minimum non dégénéré de l'énergie sur un voisinage de h de $\Phi^{-1}(y)$.

On a dans ce cas le théorème :

<u>Théorème II.2</u> ([L.5]) : <u>Il existe des réels</u> a_i $i > 0$ <u>tels que pour tout entier</u> N, <u>on a lorsque</u> $\varepsilon \to 0$

$$p_\varepsilon(x,y) = \varepsilon^{-d} \exp\left[-\frac{d^2(x,y)}{2\varepsilon^2}\right] \left(\sum_{i=0}^{N} \varepsilon^i a_i + O(\varepsilon^N)\right). \tag{2.9}$$

Preuve : Comme $\Phi^{-1}(y)$ est au voisinage de h une sous-variété, on applique la méthode des extrémas liées ([B.1], th. 1.17) ce qui permet d'éviter la transformation de Girsanov (1.8). Ensuite, on décompose l'espace de Wiener en deux comme dans le théorème II.1. Le problème consiste ensuite à montrer que la quantité intégrée $\int(\ldots)\, dP_1(vd\, w^1)$ que l'on obtient constitue une très bonne approximation de la densité $p_\varepsilon(x,y)$, dans le sens où elle en diffère de $\exp\left[-\frac{d^2(x,y)}{2\varepsilon^2}\right] O(\varepsilon^N)$ pour tout entier N. On pourrait à cette fin utiliser la méthode des grandes déviations sur le pont de J.M. Bismut ([B.1]). Dans [L.5], nous évitons de le faire, en utilisant le calcul de Malliavin ([M]). Toutefois la méthode de [B.1] donne plus d'informations. ∎

Remarque I : G. Ben Arous ([B.A.2]) obtient le même résultat grâce à la méthode de Laplace par des arguments utilisant la notion heuristique de submersion au sens faible.

Remarque II : L'exposant d dans le cas elliptique est uniquement lié au fait qu'il n'y a qu'une seule minimisante et qu'elle est non dégénérée. En particulier, si x et y sont proches, on a nécessairement ε^{-d} dans (2.9), dans le cas elliptique. La situation est ici beaucoup plus complexe : l'exposant d est aussi lié au fait que $\Phi^{-1}(y)$ ne présente pas de singularité en h. C'est ce qui rend la méthode des pas de Maslov ([Mo1], [K], [Az.1]) inopérante.

Remarque III : J.M. Bismut utilise la notion heuristique de submersion au sens fort dans sa théorie de l'indice ([B.3], [B.4], [B.5], [B.6], [B-F.1], [B-F.2]). Nous en reparlerons dans la partie suivante.

III - Utilisation de la notion heuristique de submersion au sens faible dans la théorie de l'indice :

III.1) La méthode de la chaleur et le mot "super" :

Rappelons rapidement quel est l'objet du théorème de l'indice (nous renvoyons à [Gi] et aux références y figurant pour plus de détails). Soient V une variété C^∞ compacte de dimension d et $E_\pm$ deux fibrés vectoriels hermitiens au-dessus de V. Soit D_+ un opérateur elliptique transportant les sections C^∞ de V dans E_+ sur les sections C^∞ de V dans E_-. C'est un opérateur de Fredholm. L'objet est de calculer son indice $\text{Ind}\, D_+ = \dim \text{Ker}\, D_+ - \dim \text{Coker}\, D_-$. Par toute une procédure algébrique, on se ramène au cas où V est une variété de dimension paire compacte riemanienne, spinorielle. S_+ désigne alors le fibré des spineurs de chiralité positive, et S_- celui des spineurs de chiralité négative, et l'on sait qu'il suffit de calculer l'indice de l'opérateur de Dirac D_+ tordu qui applique les sections de

$S_+ \otimes \xi$ sur les sections de $S_- \otimes \xi$ (ξ est un fibré hermitien auxiliaire). Pour plus de simplicité, nous supposerons dans toute la suite que le fibré auxiliaire ξ est réduit à 0. Soit D_- l'adjoint de D_+ : c'est aussi l'opérateur de Dirac qui applique les sections de S_- sur celles de S_+. Le principe de la méthode de la chaleur est alors le suivant :

$$\text{Ind } D_+ = \text{Tr} \exp [-\varepsilon^2 \frac{D_- D_+}{2}] - \text{Tr} \exp [-\varepsilon^2 \frac{D_+ D_-}{2}]. \tag{3.1}$$

Si $\varepsilon > 0$, $\exp [-\varepsilon^2 \frac{D_- D_+}{2}]$ et $\exp [-\varepsilon^2 \frac{D_+ D_-}{2}]$ sont représentés par des noyaux régularisants. Plus précisément, soient $h_+(y)$ une section de spineurs positives et $h_-(y)$ une section de spineurs négatifs. Il existe des opérateurs linéaires $p_\varepsilon^+(x,y)$ et $p_\varepsilon^-(x,y)$ qui dépendent de façon C^∞ de (x,y) appartenant à $V \times V$ tels que :

$$\begin{aligned} &\exp [-\varepsilon^2 \frac{D_- D_+}{2}] \, h_+(x) = \int_V p_\varepsilon^+(x,y) \, h_+(y) \, d\sigma(y) \\ &\exp [-\varepsilon^2 \frac{D_+ D_-}{2}] \, h_-(x) = \int_V p_\varepsilon^-(x,y) \, h_-(y) \, d\sigma(y), \end{aligned} \tag{3.2}$$

$d\sigma(y)$ désignant la mesure riemanienne sur V. De plus, $p_\varepsilon^+(x,y)$ est un opérateur de la fibre de S^+ au-dessus de y sur celle de S_+ au-dessus de x, et $p_\varepsilon^-(x,y)$ un opérateur de la fibre de S_- au-dessus de y sur celle de S^- au-dessus de x. La formule (3.1) nous donne :

$$\begin{aligned} \text{Ind } D_+ &= \int_V (\text{tr } p_\varepsilon^+(x,x) - \text{tr } p_\varepsilon^-(x,x)) \, d\sigma(x) = \\ &= \int_V (\text{tr}_s \, p_\varepsilon(x,x)) \, d\sigma(x) \end{aligned} \tag{3.3}$$

(tr_s pour supertrace).

L'extension de cette formule au cas d'une famille d'opérateurs de Dirac n'est pas immédiate. Donnons une approche simplifiée du problème. Soit B un ensemble de paramètres λ, ayant la structure d'une variété compacte. Supposons que V est munie d'une structure riemanienne qui dépende de façon C^∞ du paramètre $\lambda \in B$. On obtient des fibrés de spineurs S_+^λ, S_-^λ qui dépendent de façon C^∞ de λ (en schématisant), et des opérateurs de Dirac D_+^λ et D_-^λ qui dépendent de façon C^∞ de λ. Ker D_+^λ et Coker $D_+^\lambda \simeq$ Ker D_-^λ ne définissent pas des fibrés au-dessus de B, quand λ varie, car la dimension de Ker D_+^λ et celle de Ker D_-^λ varient quand λ varie. Toutefois, les sauts de dimensions de Ker D_+^λ et Ker D_-^λ sont identiques. Ceci suggère (mais cela ne fait que suggérer) que Ker D_+^λ - Ker D_-^λ constitue un fibré (virtuel ; c'est de la K-théorie ; cf. [A-S]) au-dessus de B. Nous l'appellerons le fibré indice. Lorsque l'on a un fibré (virtuel ou non), on peut lui associer un élément pair de la cohomologie paire de B : c'est son caractère de Chern. Cette association est compatible avec l'addition et la soustraction des fibrés (au sens de la K-théorie, cf. [A-S], [B.5], [Gi], [L.8]), et le caractère de Chern d'un fibré (virtuel) détermine sa classe au niveau de la K-théorie. Le problème du théorème de l'indice des familles est de calculer le caractère de Chern du fibré indice ([A-S]).

Pour obtenir une généralisation de (3.3), on peut procéder heuristiquement de la façon suivante : introduisons les espaces $H_+^{\infty,\lambda}$ et $H_-^{\infty,\lambda}$ des sections C^∞ de V sur S_+^λ et de celles de V sur S_-^λ. $H_+^{\infty,\lambda}$ et $H_-^{\infty,\lambda}$ constituent, lorsque λ varie, des fibrés de dimension infinie au-dessus de la variété des paramètres B. De plus $H_+^{\infty,\lambda}$ et $H_-^{\infty,\lambda}$ sont naturellement munis d'une structure préhilbertienne. Soient $(\mathrm{Ker}\, D_+^\lambda)^\perp$ l'orthogonal de Ker D_+^λ dans $H_+^{\infty,\lambda}$ et $(\mathrm{Ker}\, D_-^\lambda)^\perp$ celui de Ker D_-^λ dans $H_-^{\infty,\lambda}$. D_+^λ constitue un isomorphisme de $(\mathrm{Ker}\, D_+^\lambda)^\perp$ dans $(\mathrm{Ker}\, D_-^\lambda)^\perp$. On a donc "formellement" :

$$\mathrm{Ch}\,[\mathrm{Ker}\, D_+^\lambda - \mathrm{Ker}\, D_-^\lambda] = \mathrm{Ch}\,[(\mathrm{Ker}\, D_+^\lambda \oplus (\mathrm{Ker}\, D_+^\lambda)^\perp) -$$

$$- (\mathrm{Ker}\, D_-^\lambda \oplus (\mathrm{Ker}\, D_-^\lambda)^\perp)] = \mathrm{Ch}\,[H_+^{\infty,\lambda} - H_-^{\infty,\lambda}] = \tag{3.4}$$

$$= \mathrm{Ch}\,[H_+^{\infty,\lambda}] - \mathrm{Ch}\,[H_-^{\infty,\lambda}].$$

Or Ch $[H_+^{\infty,\lambda}]$ (si il existe) est un élément de la cohomologie paire de B dont le terme de degré 0 est égal à la dimension de $H_+^{\infty,\lambda}$, c'est-à-dire ici $+\infty$. (3.4) nous donne donc $+\infty - \infty$. Ceci nous incite à rechercher un formalisme qui permette de calculer le caractère de Chern de la différence de deux fibrés de dimension finie sans avoir à calculer le caractère de Chern de chaque fibré : c'est le formalisme des super-connexions de Quillen ([Q.1]). J.M. Bismut l'a étendu à la dimension infinie ([B.5], [L.8]). Il obtient ainsi une formule généralisant (3.3)

$$\mathrm{Ch}\,[\mathrm{Ker}\, D_+^\lambda - \mathrm{Ker}\, D_-^\lambda] = \int_V \mathrm{tr}_s\, p_\varepsilon^\lambda(x,x)\, d\sigma^\lambda(x). \tag{3.5}$$

Mais cette fois $p_\varepsilon^\lambda(x,y)$ est un opérateur linéaire qui applique $\Lambda(T_\lambda B) \otimes S^\lambda$ sur $\Lambda(T_\lambda B) \otimes S^\lambda$ ($S^\lambda = S_+^\lambda \oplus S_-^\lambda$), et la super-trace de $p_\varepsilon^\lambda(x,x)$ prend ses valeurs dans l'algèbre extérieure de l'espace tangent en λ à B, et non dans $\mathbb{C}$ comme dans (3.3). De plus, la formule (3.5) est valide pour toute une classe de noyau de la chaleur $p_\varepsilon^\lambda(x,y)$ sur V.

La méthode de la chaleur consiste à faire tendre $\varepsilon \to 0$ dans (3.3) et dans (3.5), en espérant que $\lim_{\varepsilon\to 0} \mathrm{tr}_s\, p_\varepsilon(x,x)$ et $\lim_{\varepsilon \to 0} \mathrm{tr}_s\, p_\varepsilon^\lambda(x,x)$ existent. Si ces deux dernières limites existent, on dit que l'on a des annulations locales des divergences. Ceci est justifié par les considérations suivantes. Prenons le cas de (3.3) pour simplifier. On sait d'après des résultats généraux d'analyse ([Gi]) qu'il existe des fonctions C^∞ de V dans $\mathbb{C}$, $c_i^+(x)$ et $c_i^-(x)$, $i \leq d$, telles que lorsque $\varepsilon \to 0$

$$(3.6)\quad \begin{aligned} \operatorname{Tr} p_\varepsilon^+(x,x) &= \frac{1}{\varepsilon^d}\Big(\sum_{i=0}^{d} c_i^+(x)\,\varepsilon^i + 0(\varepsilon^{d+1})\Big) \\ \operatorname{Tr} p_\varepsilon^-(x,x) &= \frac{1}{\varepsilon^d}\Big(\sum_{i=0}^{d} c_i^-(x)\,\varepsilon^i + 0(\varepsilon^{d+1})\Big). \end{aligned}$$

Montrer que l'on a des annulations locales des divergences revient à montrer que $c_i^+(x) = c_i^-(x)$ si $i<d$ dans (3.5), alors que l'on a toujours des annulations globales. Pour $i<d$, en effet

$$(3.6)'\quad \int_V c_i^+(x)\,d\sigma(x) = \int_V c_i^-(x)\,d\sigma(x).$$

Il ne reste plus ensuite qu'à calculer $c_d^+(x) - c_d^-(x)$ ([Gi], [B-B]).

Toutefois, M. Atiyah et E. Witten ont remarqué ([A-W]) que les annulations intervenant dans (3.6) et le calcul explicite de $c_d^+(x) - c_d^-(x)$ étaient liées à la structure de l'espace des lacets sur V. J.M. Bismut ([B.3], [B.4]) a donné un sens rigoureux aux calculs de [A-W], en utilisant une représentation stochastique convenable de $p_\varepsilon^+(x,x)$ et $p_\varepsilon^-(x,x)$ et la notion heuristique de submersion au sens fort (cf. chapitre II de cet exposé). Il a ensuite étendu sa méthode ([B.5], [B.6]) au cas de la formule (3.5). Et, il est à noter, aussi loin que nous le sachions, que sa méthode est la seule qui mette en évidence de façon naturelle la relation existant entre la structure de l'espace des lacets de V (ou de $B\times V$ pour le théorème de l'indice des familles) et les théorèmes d'indice. En particulier, sa méthode est la seule qui introduise de façon naturelle le noyau $p_\varepsilon^\lambda(x,x)$ qui permette de mener les calculs jusqu'au bout dans (3.5) (c'est-à-dire de calculer explicitement $\lim_{\varepsilon\to 0} \operatorname{Tr}_s p_\varepsilon^\lambda(x,x)$). (*)

Il y a depuis les travaux classiques répertoriés dans [Gi] d'autres méthodes pour mener les calculs jusqu'au bout, plus simples que celles de J.M. Bismut ou que celles que nous allons proposer. En analyse, par exemple, celle de E. Getzler ([Ge] pour le théorème de l'indice, [Do] pour le théorème de l'indice des familles) et celle de N. Berline et M. Vergne ([B-V.1], [B-V.2]). Celle de [Az.4] et de [P-H.1] mélange probabilité et analyse. Nous donnons ici le pendant de la méthode de J.M. Bismut en utilisant la notion heuristique de submersion au sens faible (l'idée en revient aussi à S. Watanabe et à N. Ikeda ([I-W]) qui l'ont utilisée dans la preuve de la formule de Gauss-Bonnet et celle du théorème de Hirtzebruch. Il va de soi que les calculs de [I-W] pourraient être aussi utilisés dans le contexte de [L.7] et de [L.8], et donneraient les mêmes résultats). Nous ne ferons qu'indiquer le schéma probabiliste des annulations locales apparaissant dans les théorèmes d'indices.

III.2) Le théorème d'Atiyah-Singer probabiliste :

Considérons une fonctionnelle brownienne $F(\lambda,\varepsilon,w)$ à valeurs dans $\mathbb{R}^d$, dépendant d'un paramètre $\varepsilon \in [0,1]$ et d'un paramètre λ appartenant à un ouvert K de $\mathbb{R}^p$ ou à une variété.

Nous dirons que F vérifie l'hypothèse H_1 si elle possède les propriétés suivantes :

- elle possède une version C^∞ en (λ,ε)
- elle et toutes ses dérivées en (λ,ε) sont C^∞ au sens de Malliavin
- pour tout entier j, tout multi-indice (α), la dérivée $i^{\text{ème}}$ au sens de

Malliavin de $\frac{\partial^{(j)}}{\partial\varepsilon^{(j)}} \frac{\partial^{(\alpha)}}{\partial\lambda^{(\alpha)}} F(\lambda,\varepsilon,w)$ possède une version C^∞ en (λ,ε)

- $D^{(k)}$ désignant le $k^{\text{ième}}$ gradient itéré de Malliavin, on a pour tout entier $p>0$

$$\sup_{\varepsilon\in[0,1],\ \lambda\in K} E\left[\left\| \frac{\partial^{(\alpha)}}{\partial\lambda^{(\alpha)}} \frac{\partial^{(j)}}{\partial\varepsilon^{(j)}} D^{(k)} \frac{\partial^{(\alpha')}}{\partial\lambda^{(\alpha')}} \frac{\partial^{(j')}}{\partial\varepsilon^{(j')}} F(\lambda,\varepsilon,w)\right\|^p\right] < C(p) < \infty. \tag{3.7}$$

On dira que $F(\lambda,\varepsilon,w)$ vérifie H_2 si pour tout entier $p>0$:

$$\sup_{\varepsilon\in[0,1],\ \lambda\in K} E[(\det (DF(\varepsilon,\lambda,w)^t\, DF(\varepsilon,\lambda,w)))^{-p}] < C_p < \infty. \tag{3.8}$$

Introduisons une autre fonctionnelle $G(\lambda,\varepsilon,w)$ à valeurs dans un espace vectoriel complexe de dimension finie vérifiant encore H_1. On dira qu'elle annule les divergences (condition H_3) si pour tout $j<d$

$$\frac{\partial^{(j)}}{\partial\varepsilon^{(j)}} G(\lambda,0,w) = 0. \tag{3.9}$$

Notons $\mu(\lambda,\varepsilon)$ la mesure sur $\mathbb{R}^d$ définie par :

$$f \to E[G(\lambda,\varepsilon,w)\, f(\varepsilon\, F(\lambda,\varepsilon,w))]. \tag{3.10}$$

<u>Théorème III.1</u> : <u>Supposons que</u> $F(\lambda,\varepsilon,w)$ <u>vérifie</u> H_1 <u>et</u> H_2, <u>et</u> $G(\lambda,\varepsilon,w)$ <u>vérifie</u> H_1 <u>et</u> H_3. $\mu(\lambda,\varepsilon)$ <u>possède une densité</u> $q_{\lambda,\varepsilon}(z)$ <u>lorsque</u> $\varepsilon>0$, C^∞ <u>en</u> $\varepsilon>0$, $\lambda\in K$, $z\in\mathbb{R}^d$, <u>et la loi de</u> $F(\lambda,\varepsilon,w)$ <u>possède une densité</u> $p_{\lambda,\varepsilon}(z)$ C^∞ <u>en</u> $\varepsilon\geq 0$, $\lambda\in K$, $z\in\mathbb{R}^d$. <u>De plus, on a uniformément sur</u> K :

$$\lim_{\varepsilon\to 0} q_{\lambda,\varepsilon}(0) = \frac{1}{d!} E\left[\frac{\partial^{(d)}}{\partial\varepsilon^{(d)}} G(\lambda,0,w) \,\middle|\, F(\lambda,0,w) = 0\right] p_{\lambda,0}(0). \tag{3.11}$$

<u>Preuve</u> : L'existence des densités résulte de H_1 et H_2. Quand $\varepsilon\to 0$, la densité $q_{\lambda,\varepsilon}(z)$ explose, car/la matrice de Malliavin de $\varepsilon\, F(\lambda,\varepsilon,w)$ explose. On contourne
/l'inverse/

cette difficulté en appliquant le principe de la division de ([L.2]) (cf. 1.6). On obtient comme en (1.6)

$$(3.12)\qquad \begin{aligned} q(\lambda,\varepsilon)(0) &= \lim_{n\to\infty} E[f_n(\varepsilon\, F(\lambda,\varepsilon,w))\, G(\lambda,\varepsilon,w)] = \\ &= \lim_{n\to\infty} E[f_n(F(\lambda,\varepsilon,w))\, \frac{G(\lambda,\varepsilon,w)}{\varepsilon^d}], \end{aligned}$$

f_n étant une suite de fonctions à support compact de $\mathbb{R}^d$ dans $\mathbb{R}$ tendant vers la masse de Dirac en zéro au sens des distributions. Du fait de la condition d'annulation H_3 et de la condition H_2, la mesure $\tilde{\mu}(\lambda,\varepsilon)$

$$(3.13)\qquad f \to E[f(F(\lambda,\varepsilon,w))\, \frac{G(\lambda,\varepsilon,w)}{\varepsilon^d}]$$

possède une densité C^∞ en $\lambda \in K$, <u>$\varepsilon \geq 0$</u>, et $z \in \mathbb{R}^d$.

<u>Remarque I</u> : Pour le théorème de l'indice pour un opérateur simple , $G(\lambda,\varepsilon,w)$ est à valeur dans $\mathbb{C}$. Pour le théorème de l'indice des familles, $G(\lambda,\varepsilon,w)$ est à valeurs dans une algèbre extérieure (cf. [B.3], [B.5], [L.7], [L.8]).

<u>Remarque II</u> : J.M. Bismut nous a fait remarquer que cette preuve constituait l'analogue probabiliste de la preuve de E. Getzler ([Get], [Do]).

<u>Remarque III</u> : En fait, en physique théorique ([Bo], [St]), on désire obtenir des versions beaucoup plus fines du théorème de l'indice des familles, car l'obstruction à construire une théorie consistante n'est pas de nature topologique mais locale. Un des outils les plus fréquemment utilisé est le fibré déterminant ([Q.2]) qui est un fibré de dimension 1, complexe, au-dessus de B : J.M. Bismut et D. Freed ont construit une métrique sur ce fibré, une connexion associée à cette métrique ; ils ont ensuite calculé la courbure associée à cette connexion, et l'holonomie suivant un lacet (anomalie globale) ([B-F.1], [B-F.2]). Ils utilisent dans leur traitement asymptotique la notion heuristique de submersion au sens fort. Si on voulait utiliser celle de submersion au sens faible, il faudrait utiliser des techniques semblables à celles de [I-S-T].

IV - <u>Utilisation de la notion heuristique de submersion au sens faible dans le problème de l'estimation de la densité d'une diffusion hypoelliptique lorsqu'il y a un cusp</u> :

Dans la partie II ou la partie III, nous n'avions jamais dû affronter directement le cas où Φ n'est pas une submersion au sens fort. Une des idées essentielles en la matière vient de [L.2] (on en trouve une semblable dans [St-T] en analyse), mais le traitement algébrique n'est pas encore terminé (cf. [L.6], [B-A.2],

[T]). Pour commencer, revenons à l'exemple du χ^2 à n degrés de libertés.

$F(\varepsilon,w) = \varepsilon^2 \sum_{i=1}^{n} w_i^2$. La densité de la loi de $F(\varepsilon w)$ est notée $p_\varepsilon(y)$. Soit f_m une suite de fonctions C^∞ à support compact tendant vers la masse de Dirac en 0. On a :

(4.1) $$p_\varepsilon(0) = \lim_{m\to\infty} E[f_m(\varepsilon^2 \sum_{i=1}^{n} w_i^2)] = \frac{1}{\varepsilon^2} \lim_{m\to\infty} E[f_m(\sum_{i=1}^{n} w_i^2)].$$

Ainsi contrairement à (1.6), <u>on ne doit pas diviser par</u> ε notre fonctionnelle pour faire disparaître la singularité de sa densité en 0, <u>mais par</u> ε^2. On voit alors apparaître le problème qu'il y a équivalence en dimension finie entre la notion de submersion au sens fort et celle de submersion au sens faible. Ainsi, si $m = 1$, notre χ^2 ne possède pas de densité ayant un bon comportement en 0. Comme il a été dit dans [L.6], cela suggère que les calculs en dimension infinie seront plus simples que ceux en dimension finie.

Reprenons les notations de la partie II, et supposons toujours que les champs X_i $i \neq 0$ vérifient l'hypothèse forte de Hörmander en tout point de $\mathbb{R}^d$. Comme Φ n'est pas une submersion, on ne peut appliquer la méthode des extrémas liés pour minimiser $\|h\|^2$ sur $\Phi^{-1}(y)$. Cela nous oblige à rappeler la définition d'une bicaractéristique. Soient $x \in \mathbb{R}^d$, $p \in \mathbb{R}^d$. Une bicaractéristique est la solution de l'équation différentielle sur $\mathbb{R}^d \times \mathbb{R}^d$:

(4.2) $$\begin{aligned} dx_t(x,p) &= \sum_{i=1}^{m} \langle X_i(x_t(x,p)),p_t(x,p)\rangle\, X_i(x_t(x,p))dt \\ dp_t(x,p) &= -\sum_{i=1}^{m} \langle X_i(x_t(x,p)),p_t(x,p)\rangle \\ &\quad {}^t\frac{\partial X_i}{\partial x}(x_t(x,p))\, p_t(x,p)dt \\ x_o(x,p) &= x \qquad p_o(x,p) = p. \end{aligned}$$

Posons

(4.3) $$h_{i,t}(x,p) = \langle X_i(x_t(x,p)),p_t(x,p)\rangle$$

et notons $h(x,p)$ l'élément de H_2 correspondant. La propriété essentielle pour la suite est la suivante ([B.1] 1.36) :

(4.4) $$h_{i,t}(x,p) = \langle p,(\frac{\partial x_t}{\partial x}(h(x,p)))^{-1} X_i(x_t(h(x,p)))\rangle$$

($x_t(h(x,p))$ est défini en (1.12)).

IV.1) <u>Le cas où les crochets d'ordre inférieur à deux suffisent à engendrer l'espace</u> :

Supposons maintenant que l'ensemble des crochets de Lie d'ordre inférieur à 2 construits à partir des X_i $i \neq 0$ engendrent $\mathbb{R}^d$ en le point de départ de la diffusion de l'espace. Notons $n(x,p)$ la dimension de l'image de $D\Phi(h(x,p))$ et $N(x,p)$ la quantité $n(x,p) + 2(d - n(x,p))$. On a :

Théorème III.1 [L.6] : Si p est assez petit, on a lorsque $\varepsilon \to 0$, pour tout entier N :

$$(4.5) \qquad p_\varepsilon(x,x_1(x,p)) = \varepsilon^{-N(x,p)} \exp\left[- \frac{d^2(x,x_1(x,p))}{2\varepsilon^2}\right] \left(\sum_{i=0}^{N} \varepsilon^i a_i(x,p) + O(\varepsilon^N)\right).$$

De plus, les réels $a_i(x,p)$ dépendent de façon C^∞ de (x,p) lorsque $n(x,p)$ reste constant.

Preuve : On effectue la translation $dw \to dw + \frac{h_t(x,p)dt}{\varepsilon}$, ce qui revient à introduire ([Mol], [K], [Az.1]) la solution de l'équation différentielle :

$$(4.6) \qquad \begin{aligned} dx_t(\varepsilon,dw,x,p) &= \sum_{i=1}^{m} X_i(x_t(\varepsilon,dw,x,p)) \, (\varepsilon\, dw_i + h_{i,t}(x,p)\,dt) + \\ &+ \varepsilon^2 X_o(x_t(\varepsilon,dw,x,p))\,dt \\ x_o(\varepsilon,dw,x,p) &= x. \end{aligned}$$

Soit f_n une suite de fonctions C^∞ de $\mathbb{R}^d$ dans $\mathbb{R}$, à support compact, tendant au sens des distributions vers la masse de Dirac en 0. On a δw_i, désignant la différentielle d'Ito,

$$(4.7) \qquad \begin{aligned} p_\varepsilon(x,x_1(x,p)) &= \exp\left[- \frac{\sum_{i=1}^{n} \int_0^1 h_{i,s}^2(x,p)\,ds}{2\varepsilon^2}\right] \times \\ &\times \lim_{n\to\infty} E[f_n(x_1(\varepsilon,dw,x,p) - x_1(0,dw,x,p)) \\ &\exp\left[- \frac{\sum_{i=1}^{n} \int_0^1 h_{i,t}(x,p)\,\delta w_i}{\varepsilon}\right] \end{aligned}$$

ce qui n'est rien d'autre que (1.9). Mais ici diviser par ε comme dans (1.6) ne suffit plus. On divise les composantes de $x_1(\varepsilon,dw,x,p) - x_1(0,dw,x,p)$ dans $\mathrm{Im}\, D\Phi(h(x,p))$ par ε et celles dans $(\mathrm{Im}\, D\Phi(h(x,p)))^\perp$ par ε^2. On maîtrise ensuite le terme exponentiel dans l'espérance en suivant les calculs de [D], et cela est possible lorsque p est petit. L'estimation de Varadhan permet de conclure que notre bicaractéristique est bien une minimisante.

Remarque : Lorsque l'on s'éloigne du point de départ en suivant une petite bicaractéristique, l'exposant N(x,p) décroît. Il s'agit d'un phénomène typiquement hypoelliptique dû à la présence de singularités dans $\Phi^{-1}(y)$. Il n'est pas à confondre avec les changements de puissances qui apparaissent dans le cas elliptique, qui sont dûs au fait qu'il est possible qu'il y ait un grand nombre de minimisantes reliant x à y lorsque x et y sont loins ([Mol], [Az.1], [P-H.2]).

IV.2) Etude du cas général :

Dans le cas elliptique, une minimisante est nécessairement une bicaractéristique et une petite bicaractéristique est nécessairement une minimisante de l'énergie ([Az.1]). Dans le cas hypoelliptique, J.M. Bismut ([B.1] Th. 1.17) a donné une condition suffisante pour qu'une minimisante soit une bicaractéristique. Nous allons montrer qu'une petite bicaractéristique est nécessairement une minimisante, répondant ainsi à une question de [Str].

Théorème IV.2 : Si p est assez petit, on a :

(4.8) $$d^2(x,x_1(x,p)) = \int_0^1 \sum_{i=1}^m h^2_{i,t}(x,p)\,dt = \|h(x,p)\|^2.$$

Schéma de la preuve : On a comme en (4.7)

(4.9) $$p_\varepsilon(x,x_1(x,p)) = \exp\left[-\int_0^1 \frac{\sum_{i=1}^m h^2_{i,t}(x,p)dt}{2\varepsilon^2}\right]$$

$$\bar{p}_\varepsilon(0),$$

$\bar{p}_\varepsilon(z)$ étant la densité de la mesure $\bar{\mu}_\varepsilon$.

(4.10) $$f \to E[f(x_1(\varepsilon,dw,x,p) - x_1(0,dw,x,p))$$
$$\exp\left[-\sum_{i=1}^m \int_0^1 \frac{h_{i,t}(x,p)\,\delta w_i}{\varepsilon}\right]].$$

En utilisant (4.4) et les calculs de H. Doss [D], on contrôle le terme exponentiel dans l'espérance sans difficultés si p est petit. Les calculs de [K-S.1] permettent alors de montrer que lorsque $\varepsilon \to 0$, $\bar{p}_\varepsilon(z) \le C\varepsilon^{-N}$ pour un entier N bien choisi. Cela prouve que

(4.11) $$\overline{\lim_{\varepsilon\to 0}}\; 2\varepsilon^2 \operatorname{Log} p_\varepsilon(x,x_1(x,p)) \le -\|h(x,p)\|^2.$$

De plus l'estimation de Varadhan prouve que

(4.12) $$\lim_{\varepsilon\to 0} 2\varepsilon^2 \operatorname{Log} p_\varepsilon(x,x_1(x,p)) = -d^2(x,x_1(x,p))$$

et donc $-d^2(x,x_1(x,p)) \leq -\|h(x,p)\|^2$. Comme par définition,

$d^2(x,x_1(x,p)) = \inf_{h\in\Phi^{-1}(x_1(x,p))} \|h\|^2$, on a clairement (4.8). ∎

Remarque : Supposons que $y = x_1(x,p) = x_1(x,p')$, et que p et p' sont assez petits. Dans le cas riemanien, on a $h_s(x,p) = h_s(x,p')$. Ceci reste vrai dans notre situation, si on suppose que Φ est une submersion en $h_s(x,p)$ et en $h_s(x,p')$. A cette fin, on introduit un élément g de H_2 tel que $\sum_{i=1}^m \int_0^1 g_{is}\, h_{i,s}(x,p)ds = 0$ et tel que $\sum_{i=1}^m \int_0^1 g_{is}\, h_{i,s}(x,p')ds \neq 0$, et on considère la mesure sur $\mathbb{R}^d$

$f \to E[(\sum_{i=1}^m \int_0^1 g_{is}\, \delta w_i)\; f(x_1(\varepsilon,dw))]$. On effectue le développement asymptotique de sa densité en y lorsque $\varepsilon \to 0$ de deux façons différentes : la première en faisant la translation $dw \to dw + \frac{h_s(x,p)}{\varepsilon} ds$, et la deuxième en transformant dw en $dw + \frac{h_s(x,p')}{\varepsilon} ds$. Dans le premier cas, on obtient un équivalent en $\exp[\frac{-d^2(x,y)}{2\varepsilon^2}]\, \frac{C}{\varepsilon^d}$ et dans le deuxième un terme de la forme $\exp[\frac{-d^2(x,y)}{2\varepsilon^2}]\, \frac{C'}{\varepsilon^{d+1}}$ $(C' > 0)$. D'où contradiction.

Nous allons maintenant évoquer en quelques mots la partie algébriquement difficile, en supposant que $X_o(x) = 0$ en tout x (on pourrait supposer que X_o en x appartient à l'espace engendré par les crochets construits à partir des X_i $i\neq 0$ de longueur ≤ 2). Nous renvoyons à l'exposé de G. Ben Arous dans ce volume pour plus de précisions.

Notons $F_\ell(x)$ l'espace engendré en x par les crochets de Lie construits à partir des X_i $i \neq 0$ de longueur $\leq \ell$ et posons :

(4.13) $$N(x) = \sum_\ell \ell(\dim F_\ell(x) - \dim F_{\ell-1}(x)).$$

Si $(\alpha) = (\alpha_1,\dots,\alpha_p)$, posons $|\alpha| = p$, et $X_\alpha(x) = [X_{\alpha_1} [X_{\alpha_2} \dots [X_{\alpha_{p-1}}, X_{\alpha_p}]\dots]](x)$.

<u>Théorème IV.3</u> : <u>Supposons que</u> $X_o = 0$ <u>dans</u> (2.1). <u>Il existe des constantes</u> $c_i(x)$, $c_o(x) > 0$ <u>telles que pour tout entier</u> N :

(4.14) $$p_\varepsilon(x,x) = \varepsilon^{-N(x)} (\sum_{i=0}^N c_i(x)\, \varepsilon^i + 0(\varepsilon^N)).$$

<u>De plus, si la dimension de chaque</u> $F_\ell(x)$ <u>reste constante sur une sous-variété</u> V <u>de</u> $\mathbb{R}^d$, <u>les constantes dépendent</u> $c_i(x)$ <u>dépendent de façon</u> C^∞ <u>de</u> $x \in V$.

Schéma de la preuve [L.6] : On utilise d'abord les résultats de G. Ben Arous ([B.A.4]) qui nous montrent que :

(4.15) $$x_1(\varepsilon,dw) = \exp\left[\sum_{|\alpha|\leq N} X_{(\alpha)}(x)\, \varepsilon^{|\alpha|}\, J(\alpha)\right] + O(\varepsilon^{N_1}),$$

les $J(\alpha)$ étant des expressions universelles. Il y a trop de $X_{(\alpha)}$ dans (4.15), du fait des relations de Jacobi et d'anticommutation existant dans toute algèbre de Lie. On élimine les crochets superflus et on obtient une expression du type $\exp[\Sigma X_{(\alpha)}(x)\, \varepsilon^{|\alpha|}\, \tilde{J}(\alpha)]$; ceci nous montre que la famille de variables aléatoires $\tilde{J}(\alpha)$ sur $\mathbb{R}^{\tilde{d}}$ (nous n'écrivons pas $\tilde{J}$) possède une densité $q(z)$. De plus, le calcul de Malliavin et les estimations de [K-S.2] sur le comportement quand $\varepsilon \to 0$ de la matrice de Malliavin de $x_1(\varepsilon,dw)$ montrent que l'on peut choisir N_1 assez grand dans (4.15) pour que la densité de $x_1(\varepsilon,dw)$ et celle de la loi de $\exp[\Sigma X_\alpha(x)\, \varepsilon^{|\alpha|}\, \tilde{J}(\alpha)]$ diffèrent d'au plus un $O(\varepsilon^N)$ quand $\varepsilon \to 0$. Il ne reste plus qu'à trouver un développement de cette densité : la propriété essentielle que l'on utilise est que la fonction $z_\alpha \to \exp[\Sigma X_\alpha(x)\, z_\alpha]$ est une "exponentielle". On peut donc factoriser sa première dérivée dans ses dérivées d'ordre supérieur.

Remarque I : Dans [B.A.2], G. Ben Arous obtient sensiblement les mêmes résultats par des calculs légèrement différents, en utilisant la théorie de [R-S].

Remarque II : Dans le cas où X_o est non nul, on obtient dans [L.6] théo. II.2 , un développement asymptotique de $p_\varepsilon(x,x)$ de la même forme, si on suppose que $X_o(x)$ appartient à l'espace engendré par les crochets de Lie d'ordre ≤ 2, $N(x)$ ayant une forme légèrement différente. Toutefois, on rencontre un problème majeur ([L.9]). $c_o(x)$ n'est pas forcément non nul ! (contrairement à une affirmation trop rapide dans [L.6]). De plus, si $c_o(x) = 0$, toutes les autres constantes sont nulles. Pour obtenir un contre-exemple simple, considérons le cas de $\mathbb{R}^2$ dont le point générique est noté $\begin{pmatrix} x \\ y \end{pmatrix}$, $X_o = \begin{pmatrix} 0 \\ x^2 \end{pmatrix}$, $X_1 = \begin{pmatrix} 1 \\ 0 \end{pmatrix}$, $X_2 = \begin{pmatrix} 0 \\ x^6 \end{pmatrix}$. La diffusion issue de $\begin{pmatrix} 0 \\ 0 \end{pmatrix}$ est égale à $\begin{pmatrix} \varepsilon w_1 \\ \varepsilon^4 \int_o^1 w_{1,s}^2\, ds + \varepsilon^7 \int_o^1 w_{1,s}^6\, dw_2 \end{pmatrix}$. Le $N\begin{pmatrix} 0 \\ 0 \end{pmatrix}$ théorique est égal à 5, et l'on voit bien que $p_\varepsilon\begin{pmatrix} 0 \\ 0 \end{pmatrix} = \varepsilon^{-5}\, \bar{p}_\varepsilon\begin{pmatrix} 0 \\ 0 \end{pmatrix}$, $\bar{p}_\varepsilon$ étant la densité de

$\begin{pmatrix} w_1 \\ \int_o^1 w_{1,s}^2\, ds + \varepsilon^3 \int_o^1 w_{1,s}^6\, dw_2 \end{pmatrix}$. Or $\bar{p}_o\begin{pmatrix} 0 \\ 0 \end{pmatrix}$ est la densité en $\begin{pmatrix} 0 \\ 0 \end{pmatrix}$ de $\begin{pmatrix} w_1 \\ \int_o^1 w_{1,s}^2\, ds \end{pmatrix}$

qui est nulle car $\int_0^1 w_{1,s}^2 \, ds > 0$. D'où vient le problème ?

Dans le cas de $X_o = 0$, nos $\tilde{J}(\alpha)$ sont des combinaisons linéaires d'intégrales itérées de Stratonovitch contenant autant de dw que (α) contient d'éléments. Par suite, si on remplace dans $\tilde{J}(\alpha)$ nos dw_i formellement par $h_{i,s}\, ds$, on obtient une fonctionnelle vectorielle $\tilde{J}(.)(h)$ à valeurs dans un certain $\mathbb{R}^{\tilde{d}}$ ($\tilde{d}$ étant égale au nombre de (α)). Comme nos intégrales ne contiennent pas de termes en dt, on montre que pour tout $\tilde{u}$ de $\mathbb{R}^{\tilde{d}}$, $\tilde{J}(.)$ possède la propriété suivante :

⋆ Il existe un élément $\tilde{h}$ de l'espace de Cameron-Martin tel que $\tilde{J}(.)(\tilde{h}) = \tilde{u}$ et tel que $h \to \tilde{J}(.)(h)$ en $\tilde{h}$ soit une submersion.

Dans le cas où l'on doit ajouter X_o, ceci n'est plus du tout vrai. Dans [L.9], on montre que le fait que $c_o(x) > 0$ est étroitement lié à ⋆. Ceci prouve l'importance de la matrice de Malliavin déterministe introduite par J.M. Bismut dans [B.1].

Remerciements : Nous remercions les organisateurs du colloque Franco-Japonais, M. Métivier et S. Watanabe, de nous avoir permis de publier cet exposé de synthèse sur ce qui constitue l'essentiel de notre thèse.

BIBLIOGRAPHIE

[A-S] ATIYAH M.F., SINGER I.M. : The index of elliptic operators. IV. Annals of Math. 93 (1971) pp. 119-138.

[A-W] ATIYAH M.F., WITTEN E. : Circular symmetrix and stationary phase approximation. Colloque en l'honneur de L. Schwartz 43-59 Paris Astérisque n° 131 (1985).

[Az.1] AZENCOTT R. et Alt. : Géodésiques et diffusions en temps petit. Astérisque 84-85. S.M.F. (1981).

[Az.2] AZENCOTT R. : Formule de Taylor stochastique et développement asymptotique d'intégrales de Feynman. Séminaire de Probabilités XVI (1980/81). Lecture Notes in Math. 921 Springer-Verlag.

[Az.3] AZENCOTT R. : Grandes déviations et applications. Cours de Probabilités de Saint-Flour. Lecture Notes in Math. n°774. Berlin. Springer.

[Az.4] AZENCOTT R. : Une approche probabiliste du théorème de l'indice (d'après J.M. Bismut). Séminaire Bourbaki. Exposé 633.

[B.A.1] BEN AROUS G. : Méthodes de Laplace et de la phase stationnaire sur l'espace de Wiener. Preprint.

[B.A.2] BEN AROUS G. : Développement asymptotique du noyau de la chaleur sur la diagonale. A paraître aux Annales de l'E.N.S.

[B.A.3] BEN AROUS G. : Développement asymptotique du noyau de la chaleur hypoelliptique hors du cut-locus. A paraître aux Annales de l'E.N.S.

[B.A.4] BEN AROUS G. : Flots et séries de Taylor stochastiques. A paraître dans Prob. Theory and Related Fields.

[B.1] BISMUT J.M. : Large deviations and the Malliavin calculus. Progress Math. n°45. Basel-Boston-Stuttgart Birkhäuser (1984).

[B.2] BISMUT J.M. : Mécanique aléatoire. Lecture Notes in Math n°866. Berlin Springer 1981.

[B.3] BISMUT J.M. : The Atiyah-Singer theorems, a probabilistic approach I. J. Func. Anal. 57 (1984) 56-99.

[B.4] BISMUT J.M. : Index theorem and equivariant cohomology on the loop space. Comm. Math. Physi. 98 (1985) 213-237.

[B.5] BISMUT J.M. : The Atiyah-Singer index theorem for families of Dirac operators : two heat equations proofs. Inven. Math. 83 (1986) pp. 91-151.

[B.6] BISMUT J.M. : Localization formulas, superconnections and the index theorem for families. Comm. Math. Physi. 103 (1986) 127-166.

[B-B] BOOS B., BLEECKER D.D. : Topology and Analysis. The Atiyah-Singer index formula and Gauge. Theoretic Physics. Springer.

[B-F.1] BISMUT J.M., FREED D.S. : The analysis of elliptic families I. Comm. Math. Physic. 106 (1986) 159-176.

[B-F.2] BISMUT J.M., FREED D.S. : The analysis of elliptic families II. Comm. Math. Physic. 107 (1986) 103-163.

[Bo] BOST J.B. : Fibrés déterminants, déterminants régularisés et mesures sur les espaces de modules de courbes complexes. Séminaire Bourbaki 1987. Exposé n°676.

[B-V.1] BERLINE N., VERGNE M. : A computation of the equivariant index of the Dirac operator. Bul. Soc. Math. Fr. (1985) 305-345.

[B-V.2] BERLINE N., VERGNE M. : A proof of Bismut local index theorem for a family of Dirac operators. Preprint.

[D] DOSS H. : Démonstration probabiliste de certains développements asymptotiques quasi-classiques. Bull. Sciences Mathé. 2ième série. 109 (1985) 179-208.

[Do] DONNELY H. : Local index theorem for families. Preprint.

[F] FREED D.S. : Determinants, torsion and strings. Comm. Math. Physi. 107 (1986) 483-513.

[F-S] FEFFERMAN C., SANCHEZ A. : Fundamental solutions for second order subelliptic operators. Annals of Maths. 124. 2 (1986) 247-272.

[Get] GETZLER E. : A short proof of the Atiyah-Singer index theorem. Topology 25 (1986) pp. 111-117.

[G] GAVEAU B. : Principe de moindre action, propagation de la chaleur et estimées sous-elliptiques sur certains groupes nilpotents. Acta Math. 139 (1977) 95-153.

[Gi] GILKEY P. : Tnvariance theory, the heat equation and the Atiyah-Singer index theorem. Publish. and Perish (1984).

[Gr] GREINER P. : An asymptotic expansion for the heat equation. Arch. Ratio.

Mech. Anal. 41 (1971) 163-218.

[I-W] IKEDA N., WATANABE S. : Malliavin calculus for Wiener's functional and it's application. In "From local times to global geometry". Ed. K. Elworthy Pitman (1986).

[J-S] JERISON D.S., SANCHEZ A. : Estimates for the heat kernel for the sum of squares of vector fields. Preprint.

[I-S-T] IKEDA N., SHIGEKAWA I., TANIGUCHI S. : The Malliavin calculus and long time asymptotics of certain Wiener integrals. Preprint.

[H] HÖRMANDER L. : Hypoelliptic second order differential equations. Acta Mathe. 119 (1969) 147-171.

[K] KIFER Yu I. : On the asymptotics of the transition density of process with small diffusion. Th. of Probability. Vol. 21. n°3 (1976).

[K-S.1] KUSUOKA S., STROOCK D.W. : Applications of the Malliavin calculus Part II. Preprint.

[K-S.2] KUSUOKA S., STROOCK D.W. : Applications of the Malliavin calculus Part III. Preprint.

[L.1] LEANDRE R. : Estimation en temps petit de la densité d'une diffusion hypoelliptique. CRAS. t.301. Série I. n°17 (1985).

[L.2] LEANDRE R. : Renormalisation et calcul des variations stochastiques. CRAS. t.302. Série 1. n°3. 135-138 (1986).

[L.3] LEANDRE R. : Majoration en temps petit de la densité d'une diffusion dégénérée. Z. Prob. Theo. Related fields. 74 (1987) 289-294.

[L.4] LEANDRE R. : Minoration en temps petit de la densité d'une diffusion dégénérée. A paraître aux Jour. Funct. Anal.

[L.5] LEANDRE R. : Intégration dans la fibre associée à une diffusion dégénérée. A paraître au Prob. Theor. Related Fields.

[L.6] LEANDRE R. : Développement asymptotique de la densité d'une diffusion dégénérée. A paraître in Journ. Funct. Anal.

[L.7] LEANDRE R. : Sur le théorème d'Atiyah-Singer. Preprint.

[L.8] LEANDRE R. : Sur le théorème de l'indice des familles. A paraître au séminaire de Strasbourg n°XXII.

[L-R] LEANDRE R., RUSSO F. : Estimation de Varadhan pour des diffusions à plusieurs paramètres. En préparation.

[M] MEYER P.A. : Le calcul de Malliavin et un peu de pédagogie. RCP n°25. Vol. 34. Université de Strasbourg (1984).

[Mol] MOLCHANOV S.A. : Diffusion processes and riemanian geometry. Russian Math. Surveys. 30 (1975) 1-53.

[N-S-W] NAGEL A., STEIN E.M., WAINGER S. : Balls and metrics defined by vector fields I. Acta Mathe. 155 (1985) 103-147.

[P-H.1] PEI, HSU : Preprint.

[P-H.2] PEI, HSU : Preprint.

[Q.1] QUILLEN D. : Superconnections and the Chern character. Topology 24 (1985) 89-95.

[Q.2] QUILLEN D. : Determinants of Cauchy-Rieman operators over a Riemann surface. Funct. Analy. Appli. 19 (1985) 31-34.

[R-S] ROTSCHILD L.P., STEIN E.M. : Hypoelliptic differential operators and nilpotent groups. Acta Mathe. 137. 247-320 (1976).

[S] SANCHEZ-CALLE A. : Fundamental solutions and geometry of the sum of squares of vector fields. Invent. Math. 78 (1984) 143-160.

[Sto] STORA R. : Algebraic structure of Chiral anomalies. RCP 25, Vol. 36. Université de Strasbourg (1985).

[Str] STRICHARTZ R. : Sub-riemannian geometry. J. Diff. Geometry. 24 (1986) pp. 221-263.

[St-T] STANTON K., TARTAKOFF D.S. : The heat equation for the $\bar{\partial}_b$-Laplacian. Comm. Partial Diff. Equa. 7 (1985) pp. 597-686.

[T] TOKOHOBU : En préparation.

[V] VAROPOULOS N. : Preprint.

[W] WATANABE S. : Analysis of Wiener. Functional and its applications to heat kernels. Ann. of Proba. 15. n°1 (1987) pp. 1-39.

[Da.1] DAVIES E.B. : Explicit constants for Gaussian upper bounds on heat kernels. Ameri. Journal of Maths. 109 (1987) 319-334.

[Da.2] DAVIES E.B. : Gaussian upper bounds for the heat kernels of some second order operators on riemanian manifold. A paraître au Jour. of Funct. Analysis.

[L.9] LEANDRE R. : Noyau de la chaleur hypoelliptique sur la diagonale et problème de la positivité des constantes. Preprint.

NOTE ADDED IN PRINT

<u>Bibliographie</u>: ajouter la référence:

[DUN] DUNCAN T. E. The heat equation, the Kac Formula and some Index Formula. In Partial Differential Equations and Geometry. Proceeding of the Park City Conference. Edited by C.I.Byrnes . Marcel Dekker. Feb. 1977. 19-23.

(*) les autres se contentant de <u>vérifier</u> que cette superconnexion permet de faire l'asymptotique.

ON THE FOCK SPACE REPRESENTATION OF OCCUPATIONS TIMES FOR NON REVERSIBLE MARKOV PROCESSES

Yves LE JAN
Laboratoire de Probabilités. Tour 56. Université de Paris VI
4, place Jussieu. 75252 Paris Cédex 05 (France)

The purpose of this note is first to show how the algebraic construction of the Wiener product on the Fock space can be modified to correspond to a non symmetric bilinear form on the underlying Hilbert space, secondly, to show how the expectations of such modified products can be computed as expectations of the corresponding ordinary Wiener products, with an exponential weight. These formulas can be applied to represent the occupation times of a non reversible Markov process. We shall follow the notations of our previous work [2] and omit the details in proofs similar to those given there. We refer also to [1], [3], [4], for questions of related interest.

1. The A-product.

Let $H = K + jK$ be a real Hilbert space with an isometric involution j decomposed in two orthogonal subspaces K and jK.

Let $\mathbf{H} = \bigoplus_{n,m} H^{\odot n} \otimes H^{\wedge m}$ be the associated Fock space.

j extends to an involution $u \to \bar{u}$ on $\mathbf{H}$, which reverses the $\wedge$ products.

a_x and b_x, x varying in H, are the families of bosonic and fermionic anihilation operators.

Let A be a bounded on K.

For $\mu \in K$ define

$$\Phi_\mu^A = a_\mu^* + a_{jA\mu}, \qquad \bar{\Phi}_\mu^A = a_{j\mu}^* + a_{A^*\mu},$$

$$\psi_\mu^A = b_\mu^* + b_{j\,A\mu}, \quad \bar{\psi}_\mu^A = b_{j\mu}^* - b_{A^*\mu}$$

The operators Φ_μ^A, $\bar{\Phi}_\mu^A$ commute one with each other.

$$\left[\Phi_\mu^A, \bar{\Phi}_\nu^A\right] = \left[a_\mu^*\, a_{A^*\nu}\right] + \left[a_{jA\mu}\, a_{j\nu}^*\right] = -\langle\mu, A^*\nu\rangle + \langle A\mu,\nu\rangle = 0$$

The operators $\psi^A, \bar{\psi}^A$ anticommute one with each other.

$$[\psi_\mu^A, \bar{\psi}_\nu^A]^+ = -[b_\mu^+, b_{A^*\nu}]^+ + [b_{jA\mu}, b_{j\nu}^*]^+ = -\langle\mu, A^*\nu\rangle + \langle A\mu,\nu\rangle = 0$$

To be consistent with [2], set $\Phi^I_\mu = \Phi_\mu$, $\Phi_\mu 1 = \varphi_\mu$ etc...

Proposition 1 : There exists an associative product $\times^A$ on $\mathbf{H}$ uniquely determined by the following property :

$$\varphi_\mu \times^A u = \Phi^A_\mu u, \quad \overline{\varphi}_\mu \times^A u = \overline{\Phi}^A u$$

and the same relations hold with psi's instead of phi's.

Moreover, for any $u,v \in H$, $\overline{u \times^A v} = \overline{v} \times^{A^*} \overline{u}$.

Proof : Go through the proof of Proposition 1 in [2] and make the obvious modifications.

N.B. a) In [2], $u \times^I v$ is denoted $u\,v$.

b) The A product differs of the Wick (or tensor) product $*$ only through couplings of φ with $\overline{\varphi}$ or ψ with $\overline{\psi}$.

As in [2], one checks the following formulas

$$E(\varphi_{\mu_1} \times^A \ldots \varphi_{\mu_n} \times^A \overline{\varphi}_{\nu_1} \ldots \overline{\varphi}_{\nu_m}) = \delta_{nm} \operatorname{Per}((A\mu_i, \nu_j)).$$

The permanent is replaced by a determinant if the phi's are replaced by psi's

$$E(\exp_{\times^A}(\varepsilon \Sigma \psi_{\mu_i} \times^A \overline{\psi}_{\nu_i})) = \det(I + \varepsilon M)$$

with $M_{ij} = \langle A\mu_i, \mu_j \rangle$

$$E(\psi_\mu \times^A \overline{\psi}_\nu \exp_{\times^A}(\Sigma\, \psi_{\mu_i} \times^A \overline{\psi}_{\mu_i})) = \det(C) \det\langle C^{-1} A\mu, \nu \rangle$$

with $Cx = x + \sum_1^n \langle A\mu_i, x \rangle \mu_i$

and we get similar formulas with phi's.

Note that $\lambda_{\mu \otimes \nu} \equiv \varphi_\mu * \overline{\varphi}_\nu + \overline{\psi}_\nu * \psi_\mu = \varphi_\mu \times^A \overline{\varphi}_\nu + \overline{\psi}_\nu \times^A \psi_\mu$.

Finally, $E(\psi_\mu \times^A \overline{\psi}_\nu \times^A \lambda_{\mu_1 \otimes \nu_1} \ldots \times^A \lambda_{\mu_n \otimes \nu_n}) = E(\varphi_\mu \times^A \overline{\varphi}_\nu \times^A \lambda_{\mu_1 \otimes \nu_1} \ldots)$

$$= \sum_{\sigma \in \mathcal{S}_n} \langle A\mu, \nu_{\sigma 1} \rangle \langle A\mu_{\sigma 1}, \nu_{\sigma 2} \rangle \ldots \langle \mu_{\sigma n}, \nu \rangle.$$

These last formulas will allow a representation of polynomials of the occupation times of a class of non symmetric Markov processes. We do not wish to enter into technical details but, for processes with hit points, K should contain Dirac measures and we should have $\langle A\varepsilon_x, \varepsilon_y \rangle = g(x,y)$, g being the Green function of the process. The local time at x up to ζ, L_x is represented by $\lambda_{\varepsilon_x \otimes \varepsilon_x} \equiv \lambda_x$, and

$$E_{x,y}\left(\prod_1^n L_{x_i}\right) = E\left(\psi_x \times^A \bar{\psi}_y \times^A \prod_1^n {}_{\times^A}(\lambda_{x_i})\right) \qquad \text{(see [2], II-1,2).}$$

A renormalization of the powers of the occupation field can certainly be carried out in some cases where λ_x is a generalized field, using the procedure described in [2] II-4, namely to replace the A-product by the Wick product.

N.B. If A is symmetric and positive, the A-product is simply the Wiener product associated with the new scalar product $\langle A\mu, \nu \rangle$.

2. An integration formula for A-products.

We shall now assume that A is invertible and that $B = A^{-1} - I$ is an Hilbert-Schmidt operator, either non negative and symmetric, either skew-symmetric. Consider first the case where B has finite rank :

$$B = \sum_1^N b_{ij}\, e_i \otimes e_j^*, \quad \{e_1, e_2, \ldots, e_N\} \quad \text{being an orthonormal system.}$$

Recall that the Bose field ϕ_μ can be represented by a complex gaussian field $Z_\mu = q_\mu + \sqrt{-1}\, p_\mu$.

For convenience, we shall denote Z_{e_i} by Z_i, ψ_{e_i} by ψ_i etc... .

Set $Z = \exp(\Sigma\, b_{ij}(\phi_i * \bar{\phi}_j - \psi_i * \bar{\psi}_j)) \in \bar{\mathbf{H}}$. In all cases

$$\|\Sigma\, b_{ij}\, \phi_i * \bar{\phi}_j\|^2 = \|\Sigma\, b_{ij}\, Z_i * \bar{Z}_j\|_2^2 = \|\Sigma\, b_{ij}\, \psi_i * \bar{\psi}_j\|^2 = \mathrm{Tr}(B^* B).$$

- If B is skew-symmetric, $\exp(\Sigma\, b_{ij}\, Z_i * \bar{Z}_j) = \exp(\sqrt{-1}\, \Sigma\, b_{ij}\, q_i\, p_j)$ which is a complex random variable of modulus 1. Moreover, $\|\exp(\Sigma\, b_{ij}\, \psi_i * \bar{\psi}_j)\| = \det(I + B^* B)$. Indeed, B can be put in the form

$$\Sigma\, \lambda_n (e_{2n} \otimes e_{2n+1}^* - e_{2n+1} \otimes e_{2n}^*) \quad \text{and}$$

$$\exp(\Sigma\, b_{ij}\, \psi_i * \bar{\psi}_j) = \exp(\Sigma\, \lambda_n (\psi_{2n} \bar{\psi}_{2n+1} - \psi_{2n+1} \bar{\psi}_{2n}))$$

$$= \Pi(1 + \lambda_n(\psi_{2n} \bar{\psi}_{2n+1} - \psi_{2n+1} \bar{\psi}_{2n}) - \lambda_n^2(\psi_{2n} \bar{\psi}_{2n+1} \psi_{2n+1} \bar{\psi}_{2n}))$$

$$= \Pi^*(1 + \lambda_n(\psi_{2n} * \overline{\psi}_{2n+1} - \psi_{2n+1} * \overline{\psi}_{2n}) + \lambda_n^2(1 + \psi_{2n} * \overline{\psi}_{2n}$$

$$+ \psi_{2n+1} * \overline{\psi}_{2n+1} + \psi_{2n} * \overline{\psi}_{2n+1} * \psi_{2n+1} * \overline{\psi}_{2n}))$$

the norm of which equals

$$(\Pi((1 + \lambda_n^2)^2 + 2\lambda_n^2 + 3\lambda_n^4))^{1/2} = \Pi(1 + \lambda_n^2).$$

- If B is symmetric and non negative, B can be put in the form $\Sigma \lambda_n e_n \otimes e_n^*$ with $\lambda_n \geq 0$

$$\|\exp(-\Sigma \lambda_n Z_n * \overline{Z}_n)\|_p^p = e^{p\Sigma\lambda_n} \|\exp(-\Sigma \lambda_n Z_n \overline{Z}_n))\|_p^p$$

$$= e^{p\Sigma\lambda_n} \Pi(1 + p\lambda_n)^{-1} = \det{}_2(I + pB).$$

Hence, $\|\exp(-\Sigma b_{ij} Z_i * \overline{Z}_j)\|_p^p = (\det{}_2(I + pB))^{-1}$ (cf. [5] chap. VIII).

We have also $\|\exp(\Sigma b_{ij} \psi_i * \overline{\psi}_j)\|^2 = \det{}_2(I + 2B) \det(I + 2B^2(I + 2B)^{-1})$

Indeed, $\exp(\Sigma \lambda_n \psi_n * \overline{\psi}_n) = e^{-\Sigma\lambda_n} \Pi(1 + \lambda_n \psi_n \overline{\psi}_n) = e^{-\Sigma\lambda_n} \Pi(1 + \lambda_n + \lambda_n \psi_n * \psi_n)$.

The square of its norm is $e^{-2\Sigma\lambda_n} \Pi(1 + 2\lambda_n + 2\lambda_n^2)$.

All these computations easily extend to the case where B has infinite rank, provided it is an Hilbert-Schmidt operator. Hence Z is well defined in $\overline{\mathbf{H}}$, as the product of a random variable in $\bigcap_{p>1} L_p$ and an element of $\overline{\Lambda H}$, and the formula (*) can be extended.

The formula (*) gives an alternative way to represent the occupation times for non symmetric processes, but the Hilbert Schmidt condition on the skew-symmetric part of the generator is clearly restrictive.

Remark 1 : In finite dimension, if e_i is a basis and $Z_{e_i} = Z_i = p_i + \sqrt{-1}\, q_i$, ψ_{e_i} and $\overline{\psi}_{e_i}$ can be represented by dq_i and dp_i and the Wiener product by the usual product of differential forms. "Expectations" of A-products ω^A are obtained by integration of the corresponding usual products ω under

$$\nu = \left(\frac{1}{2\pi}\right)^d \exp(\Sigma(A^{-1})_{ij}(Z_i \overline{Z}_j + dq_i \wedge dp_j)): \quad E(\omega^A) = \int_{\mathbf{R}^{2d}} \omega \wedge \nu .$$

Remark 2 : The exponentials of elements of the second chaos of $\mathbb{H}$ also allow the representation of exponentials of the occupation times L_x by the corresponding Wiener exponentials of λ_x, in the reversible case or in the case of Hilbert Schmidt perturbations. This can be extended to a large class of smooth functionals (cf. [3]).

REFERENCES :

[1] *DYNKIN, E.B.* : Polynomials of the occupation field and related random fields. J.F.A. 58, 20-52 (1984).

[2] *LE JAN, Y.* : On the Fock space representation of functionals of the occupation field and their renormalization. To appear in J.F.A.

[3] *LE JAN, Y.* : Temps local et superchamp. Séminaire de Probabilités XXI. L.N. 1247. Springer.

[4] *MEYER, P.A.* : Calculs antisymétriques et "supersymétriques" en probabilités. Séminaire de Probabilités XXII. A paraître.

[5] *NEVEU, J.* : Processus aléatoires gaussiens. Les presses de l'Université de Montréal (1968).
Additional references can be found in [2].

ON WEAK SOLUTIONS OF STOCHASTIC PARTIAL DIFFERENTIAL EQUATIONS

Michel METIVIER and Michel VIOT
Centre de Mathématiques Appliquées. Ecole Polytechnique
91128 Palaiseau Cédex (France)

Introduction.

In the seventies there was a number of works devoted to stochastic partial differential equations. After the pioneering papers of D. Dawson (1972), N. V. Krylov and B. L. Rosovski (1971), A. Bensoussan and R.Temam (1972 and 1973) appeared a number of works dealing with the strong solutions of abstract linear equations. The notion of strong solution was extended from the similar notion for stochastic Ito equations: see in particular R.Curtain (1977), M. Métivier and G. Pistone (1976).The theory of strong solutions for non linear equations was done in an extensive way by E.Pardoux (1976). The extension to stochastic partial differential equations of the notion of weak solutions of an Ito equation was made by M.Viot in his thesis (1976). But, surprisingly, while many papers have been devoted to linear equations (in particular to their so called "mild" solutions), no one seems to have dealt with weak solutions again.

The purpose of this lecture is twofold. On the one hand, we give a review of the basic theory, which is essentially Viot's theory, using recent tools for proving tightness and considering a wider class of approximation procedures (including in particular Markov chain approximations). On the other hand we go further by improving on known results regarding monotone equations.

The paper is organized as follows:
The first section recalls examples and the standard set-up for stochastic evolution equations and the notions of strong and weak solutions.

The second section presents the general scheme for studying the existence of weak solutions.
After recalling a few facts on the "energy equality" in section 3, we devote section 4 to stating and proving the existence theorem in the case of a monotone differential operator under hypotheses on the coefficients of the perturbation term, which are much weaker than the Lipschitz hypotheses used (See E. Pardoux, 1976) for the existence of strong solutions. A proper weak extension of the classical "Minty method" for deterministic partial differential equations is the essential tool.

1 The S.P.D.E. considered.

1.1. General situation

(1.1) $d\,X(t) + A\,X(t)\,dt = B\,X(t)\,dW(t)$

(1.2) **V** being a separable reflexive Banach space with dual **V'** and **H** being a separable Hilbert space identified with its dual we assume that we have the continuous dense injections $\mathbf{V}\to\mathbf{H}\to\mathbf{V}'$

(1.3) A maps **V** into **V'**

(1.4) W is a Wiener process in a Hilbert space **G**; with covariance **Q**.
$\forall\, u \in \mathbf{H}$ B(u) is a linear operator (not necessarily continuous)
from **G** into **H** , such that for every $u \in \mathbf{H}$ the operator

$$a(u) = B(u)\,\mathbf{Q}\,B^*(u)$$

is defined and is a nuclear operator in **H**.

NOTATIONS.

We shall use the following standard notations:

$(\ ;\ )$ and $|\ \ |$ will denote respectively the scalar product and the norm in **H**.

$\|\ \ \|$ will denote the norm in **V** and $\|\ \ \|_*$ the norm in **V'** , while the duality beetween **V** and **V'** will be written $\langle\ ,\ \rangle$.

1.2. Notions of Solutions

As for finite dimensional stochastic equations one has the notions of strong and of weak solutions

Strong solutions.

The basic probability space $(\Omega,\boldsymbol{F}, P)$ with its filtration $(\boldsymbol{F}_t)$ and a G-valued Wiener process W on this space beeing given, a strong solution with initial condition $x_0 \in \mathbf{H}$ is an H-valued stochastic process X such that

a) $X(.,\omega) \in L^1[O,T : V]$ and $A(X(.,\omega)) \in L^1[O,T : V']$ P.a.s.

b) $\int_{[0,t]} B(X(s))\, dW(s)$ is a square integrable **H**-valued martingale

c) $X(t) = x_0 - \int_{[0,t]} A(X(s))\, ds + \int_{[0,t]} B(X(s))\, dW(s)$

Weak solutions.

Here $(\Omega,\boldsymbol{F}, P)$ and (W_t) are not given but are part of the solution. A solution is therefore a system $(\Omega,(\boldsymbol{F}_t),P,W,X)$ where W is a G−valued Wiener process with covariance **Q** and X is a process satisfying a)b)c).

Martingale problem : M^2 (V,H,V', A,B,Q,x_0)

Let us set

$$\Omega_c := C[O,T ; H]$$

We write (C_t) for the canonical right continous filtration and (ξ_t) the canonical process on Ω_c

(i.e. $\xi_t(\omega) := \omega(t)$ and $C_t = \cap_{s>t}\sigma\{\xi_\tau \ \tau\leq t\}$)

A solution to the martingale problem M^2 (V,H,V', A,B,Q,x_0) is a probability law P on Ω_c such that

[M_1] (i) $P\{\xi_0=x_0\} = 1$

(ii) $P(\Omega_c \cap L^1[OT:V]\cap\{\omega:A(\omega)\in L^1[OT:V']\}) = 1$

[M_2] The (V'-valued) process

(1-5) $M_t= \xi_t - x_0 + \int_{[0,t]} A(\xi(s))\, ds$

takes actually its values in H and is a square integrable martingale with

(1-6) $\langle\langle M\rangle\rangle_t= \int_{[0,t]} B(\xi(s)QB^*(\xi(s))ds.$

Proposition 1

The property [M_2] is equivalent with

[M'_2] For some dense subset V_0 of V and all $\phi \in D(R)$ (the space of infinitely differentiable functions with compact support)

(i) $M_\phi^\theta(t) = \phi(<\theta,\xi(t)>) - \phi(<\theta,x_0>) - \int_{[0,t]}\phi'[<\theta,\xi(s)>]<\theta,A(\xi(s)>ds$
$-(1/2)\int_{[0,t]}\phi''(<\theta,\xi(s)>)(\theta;a(\xi(s)\theta)\, ds$

is a real martingale with

(ii) $E(\int_{[0,t]}\text{trace } a(\xi(s)ds) < \infty$

1-3. Examples.

a) <u>The Fleming-Viot Equation.</u>

This equation has been introduced as a dynamic model of the evolution of a genetic character in a population. O being a bounded domain in $\mathbb{R}^d$ and u(t,x) denoting for $x\in O$ and at time t the proportion of individuals present at time t at location x and presenting the given character. A model of the time evolution of u, taking into account migration phenomena, selection processes and the random mating of individuals is given by

$$\frac{\partial u(t,x)}{\partial t} - \Delta u(t,x) + \rho\, g(x)\, f(u(x)) = ([u(x)(1-u(x))]_+)^{\frac{1}{2}} \frac{\partial W(t)}{\partial t}$$

for $x \in \mathbf{O}$ and $t \in (0,T]$

$$\frac{\partial u(t,x)}{\partial n} = 0 \quad \text{for } x \in \partial\mathbf{O} \quad n \text{ being the inward normal at the boundary in } x$$

$$u(0,x) = u_0(x) \quad \text{with } u_0 \in L^2(\mathbf{O})$$

where W (t) is a Wiener process with nuclear covariance Q in $H = L^2(O)$ and $f: \mathbb{R} \to \mathbb{R}$ is bounded continuous.

In this example $H = L^2(O)$, V=the Sobolev space H^1 and $V' = H^{-1}$. The operaotor Δ is interpreted as the linear operator A from H^1 into H^{-1} defined by

$$\langle Au, v \rangle = -\int_{\mathbf{O}} \sum_{i=1}^{d} \frac{\partial u(x)}{\partial x_i} \frac{\partial v(x)}{\partial x_i}\, dx \qquad \text{for all u and v in } H^1$$

b) <u>A non linear equation of propagation.</u>

O is an open set in $\mathbb{R}^d$ and $p \geq 2$.

$$\frac{\partial u}{\partial t} - \sum_{i=1}^{d} \frac{\partial}{\partial x_i} \left(\left| \frac{\partial u}{\partial x_i} \right|^{p-2} \frac{\partial u}{\partial x_i} \right) = f + g(u) \frac{\partial W(t)}{\partial t}$$

$$u(o,x) = u_0(x) \quad \text{for } x \in \mathbf{O} \quad u_0 \in L^2(\mathbf{O})$$

$$u(t,x) = o \quad \text{for} \quad 0 \leq t \leq T \quad x \in \partial\mathbf{O}$$

where g is a given bounded α-Hölder continuous fonction $(0<\alpha\leq 1)$ from $\mathbb{R}$ into $\mathbb{R}$ and W is, as in example a), a Wiener process with values in $L^2(O)$ and nuclear covariance Q , whose kernel is a function q such that for some r>1

$$\int_{\mathbf{O}} |q(x,x)|^r\, dx < \infty$$

In this example $H = L^2(O)$, V is the Sobolev space $W_0^{1,p}$ while V' is its dual $W_0^{-1,p'}$ where $p' = p/(p-1)$

The existence of strong solutions of equations of such a type has being proved by E.Pardoux(1976) under Lipschitz assumptions on the mapping

$$u \to g(u(.))\, Q^{1/2}$$

assumed to be Lipschitz continuous from H into $L_2(G,H)$, the space of Hilbert-Schmidt operators from G into H.

Using Viot's ideas and more recent technics on weak compactness of processes we shall show the existence of weak solutions under weaker assumptions.

2 - Existence of weak solutions and "invariance principle".

A general way of dealing with the weak existence problem consists in building" finite dimensional" approximations X^N of the solutions, the X^N being processes taking their values in a finite dimensional subspace V_N of V, such that $\cup_N V_N$ is dense in H. The Laws P_N of these finite dimensional"approximations" can be considered as laws on C[OT;V'] or, if we consider non continuous approximations, on D[OT;V'],(the space of cad.lag V'-valued processes with its

Skorokhod topology). Then, one tries to show that they converge in some weak sense and for a suitable topology (or that a subsequence converges in some sense) to a law P on $\mathbf{D}[OT;\mathbf{V}']$, which is actually carried by $L^1[OT;\mathbf{V}] \cap C[OT;\mathbf{H}]$ and has the martingale property $[M_1]$ $[M_2]$.

2-1. Foedo-Galerkin methods would consider approximations of the following type : (e_i) being an orthonormal basis of $\mathbf{H}$ with $e_i \in \mathbf{V}$ and $\mathbf{V}_N$ beeing the subspace generated by $\{e_i : i=1\ldots n\}$ one defines for every $v' \in \mathbf{V}'$

(2-1) $\Pi_N(v') = \Sigma_{1\leq i\leq N} <e_i, v'> e_i$

For each $v \in \mathbf{V}$ we set

(2-2) $b_N(v) = \Pi_N \circ A(v)$

Then X^N would be a solution of

(2-3) $dX^N(t) = -b_N(X^N(t))\, dt + \Pi_N \circ B(X^N(t))\, d(\Pi_N \circ W(t))$

This approximation procedure has been used by Bensoussan,Teman, Métivier, Pistone, Pardoux for the study of strong solutions and by M.Viot (1976) for weak solutions.

It is to be observed that, if $u \rightarrow B(u)$ has only continuity properties, the equation (2-3) may not have the uniqueness property and that, therefore, solutions are not necessarily Markovian.

2-2. A"Random walk" approximation of (1-1)

Another type of approximation of the martingale problem [M] can be defined as follows: Let ε_N be a sequence of positive numbers converging to 0 as $N \rightarrow \infty$. For each N and each u in $\mathbf{V}_N$ we consider an orthonormal basis $\{e_i(u) : i=1\ldots N\}$ in $\mathbf{V}_N$ (considered as a subspace of $\mathbf{H}$) made of eigenvectors of the positive symmetric operator

$$a_N(u) = \Pi_N \circ B \circ \mathbf{Q} \circ B^* \circ \Pi_N(u)$$

For each N we define $\{p_i(u), q_i(u), \lambda(u) : i=1\ldots N\}$ as the solutions of

$$\lambda(u)\, \varepsilon_N^2 = \text{trace}\; a_N(u)$$

$$\lambda(u)\, \varepsilon_N^2\, (p_i(u) + q_i(u)) = (e_i(u)\, ;\, a_N(u)\, e_i(u))$$

$$\lambda(u)\, \varepsilon_N\, (p_i(u) - q_i(u)) = (b_N(u)\, ;\, e_i(u))$$

Then we consider the markov jump processes with intensity of jumps $\lambda(u)$ and with law of jumps at point u given by

$$\gamma(u, .) = \Sigma_{1\leq i\leq N}\, p_i(u)\, \delta(\varepsilon_N e_i(u)\, ,\, .) + q_i(u)\, \delta(\varepsilon_N e_i(u)\, ,\, .)$$

where $\delta(x\, ,\, .)$ is the dirac measure at point x. With the hypotheses made on the operators A and B in the following sections the law of this markov process will be uniquely defined.

Now, in the case of both approximations (2.1) and(2.2), one has defined a sequence X^N of $\mathbf{V}_N$-valued Markov processes whose Laws P_N can be considered as defined on $D[OT,\mathbf{V}]$ and have the following properties. Let us denote also by ξ the canonical process on this space, then

(2-4) $P_N(\{x_0 = P_N x_0\} \cap \mathbf{D}[OT;\, \mathbf{V}_N]) = 1$

(2-5) For every $\theta \in \mathbf{V}_N$, $\phi \in D(\mathbf{R})$ (the space of infinitely differentiable functions with compact support) the real process on $(\mathbf{D}(OT;\mathbf{V}'))$:

$M_t^\theta = <\theta, \xi(t)> - <\theta, \xi_0> - \int_{]0,t]} <\theta, A(x_s)>\, ds$

is a martingale with

$< M_t^\theta > = \int_{]0,t]} <\theta;\, a(\xi(s))\theta)\, ds$

Or, equivalently

(2-5') For every $\theta \in V_N$, $\phi \in D(\mathbf{R})$

$$M_\phi^\theta(t) = \phi(<\theta,\xi(t)>) - \phi(<\theta,x_o>) + \int_{]0,t]} \phi'(<\theta,\xi(s)>) <\theta;A(\xi(s))> ds$$
$$-1/2 \int_{]0,t]} \phi''(<\theta,\xi(s)>) <\theta;a(\xi(s)\theta)> ds$$

is a P^N- martingale.

2-3. First scheme of proof of the existence theorem.

From these relations it is clear that one possible scheme of proof of the existence of a weak solution is the following .

Show that the P^N are carried by a subspace Ω^* of $D[OT;V'] \cap L^p[OT;V]$, which can be endowed with a Lusin topology $\mathcal{T}$ such that

a) The P^N are tight for $\mathcal{T}$.
b) For any weak limit P of (P^N) the functionals

$\omega \to M_\phi^\theta(t,\omega)$ are P.a.s. continuous for $\mathcal{T}$.

c) $P(\Omega^* \cap C[OT,H]) = 1$

This scheme is the one used by M.Viot to deal with the case of compact injections $V \to H \to V'$ and $A: u \to A(u)$ continuous bounded from V into V' and "coercive".

If $A: V \to V'$ is not assumed to be continuous (in example 2 above A is only "hemicontinuous"), a topology $\mathcal{T}$ with the above property may not be easily available. In particular a topology $\mathcal{T}$ for which (P_N) is tight and

$\omega \to \int_{]0,t]} < \theta, A\omega(s) > ds$

is also continuous.

To circumvent this difficulty Viot proposed, when A is monotone, an extension of the Minty method.This is what we shall carry out completely later in order to get an existence theorem under fairly weak assumptions on the continuity of coefficients.

3 - Energy equality and weak compactness of the laws (P_N)

3 .1. A priori Bounds.

Proposition 2.

Let us assume that the operator A has the following set [I] of properties (for some $p \geq 2$).

$[I_a]$ **Coercivity.** There exist constants $\gamma, \nu \geq 0$ and $\alpha > 0$ such that for every $u \in V$

(3-1) $<u, Au> + \gamma |u|^2 + \nu \geq \alpha \|u\|^p$

$[I_b]$ **Boundedness.** There exists a constant K such that for every $u \in V$

(3-2) $\|Au\|_* \leq K(1 + \|u\|^{p-1})$

Let us also make the following assumption on B :

$[I_c]$ For every $u \in H$

(3-3) $\operatorname{trace} B(u) Q B^*(u) \leq K(1 + |u|^2)$

for some constant K.

Then the probability laws (P_N) in $D[OT;V']$ have the following properties .

(3-4) $\sup_N E_N (\sup_{t \leq T} \|\xi_t\|^2) \leq K_T(1 + |x_0|^2)$

(3-5) $\sup_N E_N \int_{]0,t]} \|\xi_s\|^p ds \leq K_T(1+ |x_0|^2$

Proof.

The process ξ being P_N-a.s. finite dimensional, we may apply the usual Ito formula to get

$$|\xi_t|^2 = |x_0|^2 - 2\int_{]0,t]} \langle \xi_s, A(\xi_s) \rangle ds + 2\int_{]0,t]} (\xi_s; dM_s) + [M]_t$$

where M is a local martingale with quadratic variation [M].

The coercivity of A then gives

(3-6) $|\xi_t|^2 + \alpha \int_{]0,t]} \|\xi_s\|^p \, ds = |x_0|^2 + 2\gamma \int_{]0,t]} |\xi_s|^2 \, ds + \nu t + [M]_t + 2 \int_{]0,t]} (\xi_s ; dM_s)$

and therefore

$$|\xi_t|^2 \le |x_0|^2 + 2\gamma \int_{]0,t]} |\xi_s|^2 \, ds + \nu t + [M]_t + 2 \int_{]0,t]} (\xi_s ; dM_s)$$

Taking expectation on both sides, we obtain, for every stopping time τ which localizes $((\xi_t ; M_t))_{t\ge 0}$:

$E_N(|\xi_{t\wedge\tau}|^2) \le E_N(|x_0|^2) + 2\gamma \int_{]0,t\wedge\tau]} E_N(|\xi_s|^2) \, ds + \nu t + \int_{]0,t\wedge\tau]} E_N (\text{trace } a(\xi_s)) \, ds$

Using hypothesis $[I_c]$ and the Gronwall inequality one easily derives from the later inequality the conclusion (3-4) of Proposition 2.

Using again (3-6), $[I_c]$ and Inequality (3-4) one obtains (3-5)

3-3 Compactness properties.

Let us set

$$\Omega^* := \mathbf{D}[OT;V'] \cap L^\infty[OT;\mathbf{H}] \cap L^p [OT;\mathbf{V}]$$

We observe that $\mathbf{D}[OT;\mathbf{H}_\sigma] \supset \mathbf{D}[OT;V'] \cap L^\infty[OT;\mathbf{H}]$

We define the following topologies on Ω^*:

T_1:= Topology induced by the Skorokhod topology of $\mathbf{D}[OT;V']$

T_2:= Topology induced by the weak topology of $L^p [OT;\mathbf{V}]$

T_3:= Topology induced by $L^2[OT;\mathbf{H}]$

T_4:= Topology induced by $\mathbf{D}[OT;\mathbf{H}_\sigma]$.

Remark.

From classical measurability properties in separable locally convex spaces, Ω^* is a Borel subset of any one of the spaces $\mathbf{D}[OT;V']$, $L^\infty[OT;\mathbf{H}]$, $L^p [OT;\mathbf{V}]$ etc.

As such the topologies of these spaces induce a Lusin topology on Ω^*. (See for details L. Schwartz, 1973, or A. Badrikian, 1970).

Proposition 3.

Under the hypotheses of proposition 2 the laws P_N are carried by Ω^* and are tight for the Lusin topology $T = T_2 \vee T_4$

Proof

Since Ω^* is a Borel subset of $\mathbf{D}[OT;\mathbf{H}_\sigma] \cap L^p [OT;\mathbf{V}]_\sigma$, it is enough to show that the (P_N) are tight on $\mathbf{D}[OT;\mathbf{H}_\sigma]$ for T_4 and on $(L^p [OT;\mathbf{V}])_\sigma$.

The tightness for T_2 is a trivial consequence of (3-5), while the tightness for T_4 can be obtained relatively easily in the following way : use the decomposition

$$\text{(3-7)} \qquad \xi_t = \Pi_N x_o - \int_{[o,t]} b_N(\xi_s) \, ds + M_t$$

with

$$\text{(3-8)} \qquad \langle\langle M \rangle\rangle_t = \int_{[o,t]} a_N(\xi_s) \, ds \qquad \text{for the law } P_N.$$

We observe from (3-4) that the tightness of the marginals $(\xi_t^{-1} \circ P_N)$ for every t is trivial.

From a Rebolledo-type sufficient condition for tightness (see for example R. Rebolledo, 1979, or A. Joffe and M. Métivier, 1986), it is enough to show that, for every $\theta \in \mathbf{H}$, $\varepsilon > 0$, there exist δ such that for any family $(\tau (N))$ of stopping times

$$\text{Sup}_N \, | E^N(\int_{]\tau (N), \tau (N)+\delta]} (\theta ; b^N(\xi_s)) \, ds) | \le \varepsilon$$

$$\text{Sup}_N \, | E^N (\int_{]\tau (N), \tau (N)+\delta]} (\theta ; a^N(\xi_s)\theta) \, ds) | \le \varepsilon$$

This reads: for every $\theta \in \cup_N \mathbf{V}_N$

$$(3.9)\ \mathrm{Sup}_N\ |E^N(\int_{]\tau(N),\tau(N)+\delta]} \langle\theta, A(\xi_s)\rangle\, ds)| = \mathrm{Sup}_N\ |E^N(E^N{}_{\xi(\tau(N))}(\int_{[0,\delta]} (\theta ; A(\xi_s))\, ds))| \le \varepsilon$$

and

$$(3.10)\ \mathrm{Sup}_N\ |E^N(\int_{]\tau(N),\tau(N)+\delta]} <\theta, a(\xi_s)\theta>\, ds)| = \mathrm{Sup}_N\ |E^N(E^N{}_{\xi(\tau(N))}(\int_{[0,\delta]} (\theta ; a(\xi_s))\, ds)| \le \varepsilon$$

But these majorations follow easily from the a priori bounds of proposition 2 .

Proposition 4.

If to the hypothesis of Proposition 3 we add :

$[II_a]$ Compactness Hypothesis : the injections $V \to H \to V'$ are compact

$[II_b]$ Existence of a special basis : there exists an orthonormal basis (e_i) of H included in V such that the projection operators Π_N defined by (2-1) are such that

$$\mathrm{Sup}_N \|\Pi_N\|_{L(V',V')} < \infty$$

Then the laws (P^N) form a tight family on Ω^* for the topology $T^* = T_1 \vee T_2 \vee T_3$.

Proof

$[II_a]$ and $[II_b]$ allow to prove the tightness in $D[OT,V']$. In fact in view of $[II_b]$ one may replace $|\langle\theta, A(\xi_s)\rangle|$ by $\|A(\xi_s)\|_*$ and $|<\theta, a(\xi_s)\theta>|$ by trace $a(\xi_s)$ in the inequalities (3.9) and (3.10) above.This gives a Rebolledo condition for the V'-valued process ξ, while $[II_a]$ and the inequality (3.4) give the tightness in V' of

the laws $(\xi_t^{-1} \circ P_N)_N$. This proves the tightness in $D[OT,V']$. Moreover a classical compactness theorem in fonctional analysis tells that if $H \to V'$ is compact, any subset of $L^2[OT,V]$ which is bounded in $L^2[OT,V]$ and compact as a subset of $L^2[OT;V']$ is also relatively compact in $L^2[OT,H]$. If we then consider a set $K_1 \cap K_2$ with K_1 bounded in $L^2[OT,V]$, K_2 compact in $D[OT,V']$ and therefore in $L^2[OT,V']$, $K_1 \cap K_2$ is relatively compact in $L^2[OT,H]$. The tightness for T_1 and T_2 implies then immediately the tightness for T_3 and therefore the proposition.

3-4 Energy equality

We say that we have the energy equality for a mesure P on $D[OT,V'] \cap L^1[OT,V]$ if, as soon as

$$\xi_t = x_0 - \int_{]0,t]} A(\xi(s))\, ds + M_t$$

with M_t a right-continuous H-valued martingale, the following equality holds

$$(3\text{-}11)\quad |\xi_t|^2 = |x_0|^2 - 2\int_{]0,t]} <\xi_s, A(\xi(s))\, ds + 2\int_{]0,t]} <\xi_s, dM_s> + [M]_t$$

Proposition 5 - (Viot-Pardoux)

The energy equality holds for every P, in particular, in the following two cases

- There exists a "special basis "of H.
- V' is strictly convex (or more generally there exists a monotone coercive hemicontinuous operator A from V into V' : cf E. Pardoux 1976).

Consequence

If (3-11)) holds with a martingale M which is continuous, then the paths of $|\xi|^2$ are a.s. continuous. Therefore it is enough to know that the paths are a.s. continuous in $C[OT;H_\sigma]$ to conclude they are in $C[OT;H]$ a.s.

4. Existence theorems without monotony assumption

4-1 Remark on the linear case (i.e. :A linear continuous $V \to V'$).

In the linear case the existence theorem is deduced in a rather trivial way from Proposition 4.

In this case the functional

$$\omega \to \int_{]0,t]} \phi' (< \theta, \xi(s,,\omega)) >) < \theta, A\ \xi(s,\omega) > ds$$

is continuous for the topology $T_2 \vee T_4$.

Let us add the assumption

[III] For every $\theta \in \cup_N V_N$ the mapping $u \to (\theta; a(u)\theta)$ is continuous on H_σ (H endowed with the weak topology).

Then the functional

$$\omega \to \int_{]0,t]} \phi'' (< \theta, \xi(s,\omega)) >) (\theta; a\ \xi(s,\omega)\ \theta)\ ds$$

is continuous on Ω^* for the topology T_4. Applying the method exposed in 2.3 we immediately obtain the following statement.

Theorem 1.

If A is linear continuous from V into V' the hypotheses [I],[III] are sufficient for the existence of a weak solution for (1-1) .

More generally, for every A such that $\omega \to < \theta, A\ \xi(s,\omega) >$ is continuous on $(L^p([OT,V]_\sigma)$, the same conclusion holds.

Example

Let $H = L^2(O)$, O being an open subset of $\mathbb{R}^d$, and let W be a brownian motion with values in **H** with covariance defined by a kernel $q(.,.) \in L^\infty(O \times O)$. L et f be a continuous function on $\mathbb{R}$ such that $|f(u)| \leq (1+|u|)$. If B(u) is for every $u \in H$ the (unbounded in H) operator from $L^\infty(O)$ into $L^2(O)$ defined by $B(u)\ h := f(u)h$, it is easily seen that the hypothesis [III] holds.

4-2 Case of a special basis and compactness.

In the case where the hypotheses [II] hold, the sequence (P_N) is also tight for

$T^* = T_1 \vee T_2 \vee T_3$

we can therefore afford hypotheses on A and a which make

$$\omega \to <\theta, A\xi(\bullet,\omega)>$$

and

$$\omega \to (\theta', a(\xi(\bullet,\omega))\theta)$$

continuous for this topology. Since the following hypotheses on A are enough to guarantee that

$\omega \to (\theta; A_2\xi(\bullet,\omega))$ is continuous on $L^2[OT;H]$, we may then state the following

Theorem 2. (Viot)

Let us assume [I] and [II] and also

[III]' $u \to (\theta; a(u)\theta)$ continuous on **H**.

[IV] $A = A_1 + A_2$ where

A_1 linear continuous from V into V'.

$A_2 \in C(H;H_\sigma)$ with $|A_2(u)| \leq K(1+|u|)$.

Then the equation (1-1) has a weak solution.

5. Case A monotone.

We consider the case of an A with the properties

$[V_a]$ (Monotony) : $\exists \lambda \geq 0$ such that for every $u, v \in V$

$$< u-v , A(u) - A(v) > + \lambda\ |u-v|^2 \geq 0$$

$[V_b]$ ("strong" hemicontinuity) : Restricted to every finite dimensional subspace V_0 of V the mapping $v \to Av$ is continuous from V_0 into V'.

In this case, because of the lack of continuity of the functionals $M_t\phi$ one cannot immediately conclude to the martingale property of any limit P of the sequence (P_N).

One set

$$\mathcal{V} := L^p[OT;V]$$

$$\mathcal{V}' := L^{p'}[OT;V']$$

Let us call $\mathcal{V}'_\sigma$ (resp. $\mathcal{V}_\sigma$) the space $L^{p'}[OT;V']$ (resp. $L^p[OT;V]$) endowed with the weak topology. One introduces the laws P^*_N on

$\Omega^{**}=\Omega^* \times \mathcal{V}'_\sigma$, image of the laws P_N by the mapping $\omega \to (\omega, A\omega)$.

We call $\qquad \xi(t, \omega \times v') := \xi(t,\omega)$

$\qquad \chi(t,\omega,v') := v'(t)$

On Ω^{**} one has the natural right continuous filtration generated by ξ^* and χ.
The (P^{*N}) have the property

(i) $\quad P^{*N}\{(\omega;v') : A(\omega) = v'\} = 1$

(ii) $\quad$ The process M on $\Omega^* \times \mathcal{V}'_\sigma$ defined by

$$M_t(\omega,v') := \xi_t(\omega,v') - \xi_0(\omega,v') + \int_{]0,t]} v'(s)\, ds$$

is an $\mathbf{H}$-valued right continuous square integrable P^{*N}-martingale with

$$\langle\langle M \rangle\rangle_t = \int_{]0,t]} a(\xi(s))\, ds.$$

5-1 A Minty stochastic Lemma.

Lemma

Let Ω^* be a borel subset of $D[OT,V'] \cap L^p[OT;V] \cap L^\infty[OT;\mathbf{H}]$ endowed with a Lusin topology $\mathcal{T}$ finer than the topology induced by $\mathbf{D}[OT;\mathbf{H}_\sigma]$. Let (P^{*N}) be a sequence of probability measures on $\Omega^{**} = \Omega^* \times \mathcal{V}'_\sigma$, which form a tight sequence on $\Omega^* \otimes \mathcal{V}'_\sigma$ and has the properties (i) and (ii). Let P^* be a weak limit of (P^{*N}) such that (ii) holds for P^* and carried by $C[OT;\mathbf{H}] \times \mathcal{V}'_\sigma$. Then, assuming [V] for A and assuming also that the mapping $u \to$ trace $a(u)$ has continuity properties insuring the continuity of

$$\omega \to \int_{]0,T]} \text{trace}\,(a(\xi(s,\omega))\, ds$$

for the topology $\mathcal{T}$, the limit P^* has also the property (i).

Proof

To simplify the exposition we give the proof in the case $\lambda = o$. We prove first the following property : for every measurable

$$V: \Omega \times \mathcal{V}'_\sigma \to \mathcal{V}_\sigma \text{ such that}$$

$$\int P^*(d\omega,dv') \int_{[0,T]} \| V_t(\omega,v') \|^p dt < \infty$$

one has

(5-1) $\quad \int P^*(d\omega,dv') \int_{[0,T]} \langle \xi(s,\omega) - V_s(\omega,v'), \chi(s,v') - A(V_s(\omega,v')) \rangle\, ds \geq 0$

If such an inequality holds, setting

$$V_s(\omega,v') = \xi(s,\omega) - \rho U_s(\omega,v')$$

for $\rho > o$ and U any bounded measurable mapping from $\Omega^* \times \mathcal{V}'_\sigma$ into $\mathcal{V}_\sigma$ one obtains

$$\int P^*(d\omega, dv) \int_{[0,T]} \langle U_s(\omega,v'), \chi(s,v') - A(\xi(s,\omega) - \rho U_s(\omega,v')) \rangle\, ds \geq 0$$

But, because of the hemicontinuity, the expression

$$\langle U_s, \chi_s - A(\xi_s - \rho U_s) \rangle \text{ tends to } \langle U_s, \chi_s - A(\xi_s) \rangle \text{ a.s.}$$

with

$$|<U_s, \xi_s - A(\xi_s - \rho U_s)>| \leq K \ ||U_2|| \ [1 + (||\xi_s|| + ||U_s||)\ ]^{p-1/p}$$

We can therefore pass over to the limit in the above inequality , which gives

$$\int P^*(d\omega, dv') \int_{]0,t]} < U_s(\omega,v') , \chi_s(v') - A(\xi_s(\omega))> ds \geq 0$$

Since this holds for every U bounded mesurable , one has

$$\chi_s(v') = A\, \xi_s(\omega) \qquad P^*.a.s.$$

To prove (5-1) we first observe that the continuous functions V of the form

$$V_t(\omega,v') = \Sigma_{1\leq i\leq k}\, \phi_t^i(\omega,v')\, v_i \qquad v_i \in V$$

form a dense set in $L^p(\Omega^{**},P^*,\nu)$. We restrict therefore ourselves to functions $V=\phi(\omega,v')v_o$, $v_o \in V$ with ϕ a real continuous process with paths in $L^p[0,T]$.

We set

$$\Psi(\omega,v') = \int_{]0,T]} < \xi_s(\omega) - V_s(\omega,v') , \chi_s(v') - A(V_s(\omega,v')) > ds$$

and observe that in view of (i) and the monotony property of A

$$\int \Psi(\omega,v')\, P^*_N(d\omega,dv') \geq 0$$

We decompose

$$\Psi(\omega,v') = \Psi_1(\omega,v') + \Psi_2(\omega,v')$$

$$\Psi_1(\omega,v') = \int_{]0,T]}\{< \xi_s(\omega), \chi_s(v')> - (1/2) \text{ trace } a(\xi(s))\}\, ds$$

$$\Psi_2(\omega,v') = -\int_{]0,T]}\{ < \xi_s(\omega), A(\phi_s(\omega,v')v_o) > + < V_s, \chi_s - A(V_s) > + (1/2)\text{trace } a(\xi_s)\} ds$$

Because of the hypotheses made on V and a Ψ_2 is continuous.

Ψ_1 is not continuous, but, because of the property (ii) for P^* and P^{*N} and the energy-equality, if we set

$$\Psi'_1(\omega,v') = (1/2)\, (|x_o|^2 - |\xi_t|^2)$$

we have

$$E^*\, \Psi_1(\omega,v') = E^*\, \Psi'_1(\omega,v') \qquad \text{and} \qquad E^{*N}\, \Psi_1(\omega,v') = E^{*N}\, \Psi'_1(\omega,v')$$

But Ψ'_1 is upper semicontinuous on $C[OT;H_\sigma]$.

Therefore

$$0 \leq \lim_N \sup \int \Psi(\omega,v')\, P^{*N}(d\omega,dv') = \lim_N \sup \int (\Psi_1 + \Psi_2)\, dP^{*N}$$

$$= \lim_N \sup \int (\Psi'_1 + \Psi_2)\, dP^{*N} \leq \int (\Psi'_1 + \Psi_2) d\, P^*$$

$$\leq \int (\Psi'_1 + \Psi_2) d\, P^* = \int \Psi(\omega,v')\, P^*(d\omega,dv')$$

This shows (5.1), and therefore the lemma.

Theorem 3

a) Under the hypotheses [I] ,[II], [V] and a condition on $u \to$ trace $a(u)$ insuring the continuity of $\omega \to \int_{[0,T]}$trace $a(\xi_s(\omega))\, ds$ for $\mathcal{T}^* = \mathcal{T}_1 \vee \mathcal{T}_2 \vee \mathcal{T}_3$

The equation (1-1) has a weak solution .

b) Under the hypothesis [I,], [V] + [III] on the mapping

$$\omega \to \int_{[0,T]} \text{trace } a(\xi_s(\omega))\, ds \quad \text{for } \mathcal{T} = \mathcal{T}_2 \vee \mathcal{T}_4$$

The equation (1-1) has a weak solution .

Proof.

Under the set a) ((resp.b)) of hypotheses, we may apply the above Minty Lemma, which gives the martingale property in the form $[M_2]$ for any weak limit P^* of the sequence P^{*N} of Probability measures on the Lusin space $(\Omega^*, \mathcal{T}^*)$ (resp$(\Omega^*, \mathcal{T})$).

Final remark

If one has the weak uniqueness property for the equation (1-1) this gives naturally an "invariance principle". Some uniqueness results can be found in M. Viot (1976) and M. Métivier (1987), which extend some uniqueness properties for Ito equations as proved, for example, in T. Yamada and S. Watanabe (1971)

Bibliography

A. BADRIKIAN, Séminaire sur les fonctions aléatoires linéaires et les mesures cylindriques, *Lect. Notes in Math.* n° 139, Springer Verlag 1970.

A. BENSOUSSAN, R. TEMAM. [1] Equations aux dérivées partielles stochastiques non linéaires. *Israel J. of Math.* vol. II n°1, 1972 pp.95-129.

[2] Equations stochastiques du type Navier-Stokes , *J . Funct. Anal.* 13, 2 (1973).

P. BILLINGSLEY. *Convergence of probability measures.* Wiley, New York (1969).

N. BOURBAKI. *Integration* chap. IX, Hermann, Paris 1969.

R.F. CURTAIN. Stochastic evolution equations with general white noise disturbance. J. Math. Anal. Appl., 60, pp 570-595 (1977).

R.F. CURTAIN, P. L. FALB. Stochastic differential equations in Hilbert spaces. *J. Differential equations*, 10, pp 412-430 (1971).

G. DA PRATO, M. IANNELLI, L. TUBARO. Some results on linear stochastic differential equations in Hilbert spaces. *Sochastics, 6, pp 105-116 (1982).*

D.A. DAWSON. Stochastic Evolution Equations, *Math. Biosciences* 15, 287-316 (1972).

W.H. FLEMING.[1] Distributed parameter stochastic systems in population biology. pp.179-191. Control Theory, Numerical Methods and computer Systems Modelling.*Lectures Notes in Economies and Mathematical Systems* n° 107.Springer Verlag (1975)

[2] C.H. SU, Some one-dimensional Migration Models in Population Genetic Theory, Theoritical Population Biology, 5, 431-449, (1974).

A. ICHIKAWA. Linear stochastic evolution equations in Hilbert spaces. *J. Diff. Equat.* 28, pp.266-283 (1978).

N. IKEDA, S. WATANABE. *Stochastic differential equations and Diffusion Processes.* North-Holland. Amsterdam (1981).

K. ITO. [1] Infinite dimensional Ornstein-Uhlenbeck processes, *Proc. Taniguchi International Conference on Stochastic Analysis,* 1982, Katata-Tokyo, 1984.

[2]*Foundations of Stochastic Differential Equations in Infinite Dimensional Spaces.* CBMS-NSF Regional conference series in Applied Mathematics n°47.SIAM. Philadelphia. 1984.

J. JACOD, M. METIVIER, J. MEMIN. On tightness and stopping times. *Stoch. Proc. Appl.* **14** (1983) pp 109-141

A. JOFFE, M. METIVIER, Weak convergence of sequences of semimartingales with Applications to Multitype branching processes. *Advances in Applied Probability* , **18**, pp 20-65,1986.

N.V. KRYLOV, B.L. ROSOVSKI. Cauchy problem for linear stochastic partial differential equations. *Izvestia Akademia Nauk CCCR. Math. series.* 41 (6), pp 1329-1347. (1971).

J.L. LIONS, *Quelques méthodes de résolution des problèmes aux limites non linéaires* , Dunod, Gauthier villars , Paris 1969.

J.L. LIONS, E. MAGENES, *Problèmes aux limites non homogènes et applications* , vol. I, Dunod (1968).

M. METIVIER, [1] *Semimartingales* . De Gruyter. Berlin, New York, 1982.

[2] *Weak convergence of processes. Infinite dimensional Invariance Principles. Weak solutions of stochastic partial differential equations.* Course: Scuola Normale Superiore di Pisa. 1987.

M. METIVIER, J. PELLAUMAIL. *Stochastic Integration.* Academic Press. New York.(1980)

M. METIVIER, G. PISTONE.[1]. Sur une équation d'évolution stochastique. *Bull. Soc. Math. France.* **104.** (1976) pp 65-85.

[2] Une formule d'isométrie pour l'intégrale stochastique hilbertienne et équations d'évolution linéaires stochastiques. Z.W. **33** (1975) pp 1-18

E. PARDOUX . *Equations aux dérivées partielles stochastiques non linéaires monotones. Etude des solutions fortes de type Ito.* Thèse Université de Paris Sud. Orsay, Novembre 1975.

K.R. PARTHASARATHY, *Probability measures on metric spaces* . pp 132-150. Academic Press. 1967.

P. PRIOURET, Ecole d'été de Saint-Flour III. *Lect. notes in Math* n° 390, Springer Verlag (1974).

R. REBOLLEDO, La méthode des martingales appliquée à la convergence en loi des processus. *Mémoires de la S. M. F.*,t 62 , 1979.

L. SCHWARTZ, [1] *Séminaire Ecole Polytechnique* 1969-1970 : Applications radonifiantes .
[2]*Random measures on arbitrary topological spaces and cylindrical measures* Oxford Univ. Press. 1973.

D.W.STROOCK - S.R.S.VARADHAN. [1] Diffusion Processes with continuous coefficients. *Com. pure and appl. Math.* (22) pp. 345-400 and 479-530, (1969).
[2]*Multidimensional diffusion processes*. Springer.Berlin-Heidelberg-New York. (1979)

M. VIOT. [1] *Solutions faibles d'équations aux dérivées partielles non linéaires*. Thèse Université Pierre et Marie Curie ; Paris. (1976).
[2] Solution en loi d'une équation aux dérivées partielles stochastiques non linéaire: Méthodes de compacité. *Comptes-rendus Acad.Sciences*. t.278 A (1974) pp1185-1188.
[3] Solution en loi d'une équation aux dérivées partielles stochastique non linéaire: méthode de monotonie. *Comptes-rendus Acad.Sciences*. t.278 A (1974) pp 1405-1408.

J. WALSH. *An Introduction to Stochastic Partial Differential equations*. Lectures Notes in Maths.n° .Springer.

T. YAMADA, S. WATANABE, On the uniqueness of solutions of stochastic differential equations. J. *Math. Kyoto Univ.* vol. 11, pp. 155-167 et 553-563 (1971).

NOTE ADDED IN PRINT

Formula (2.5') Read:

(2-5') For every $\theta \in \mathbf{V}_N$, $\phi \in \mathcal{D}(\mathbf{R})$ $\exists$ un processus $R_\phi{}^\theta(N,t)$ et une constante $K(\phi,t)$ with $|R_\phi{}^\theta(N,t)| \leq K(\phi,t)\varepsilon_n$, such that

$$M_\phi{}^\theta(t) = \phi(\theta <\theta,\xi(t)>) - \phi(<\theta,x_0>) + \int_{]0,t]} \phi'(< \theta,\xi(s)>) <\theta;A(\xi(s))> ds$$

$$- 1/2 \int_{]0,t]} \phi''(< \theta,\xi(s)>) <\theta;a(\xi(s)\theta)> ds - R_\phi{}^\theta(N,t)$$

is a P^N- martingale.

Formula (3.4) Read:

(3-4) $\sup_N E_N (\sup_{t\leq T} |\xi_t|^2) \leq K_T(1 + |x_0|^2)$

UNE REMARQUE SUR LES CHAOS DE WIENER

Paul-André MEYER
Institut de Recherche Mathématique Avancée.
Rue du Général Zimmer. 67084 Strasbourg Cédex (France)

Les résultats présentés dans cette note ont été obtenus en commun avec Y.Z. Hu , en essayant de comprendre certains aspects de l'intégrale de Feynman. Nous ne dirons rien ici de cette motivation, qui est présentée dans un article (d'exposition) à paraître dans le volume XXII du Séminaire de Probabilités, dans la série des Lecture Notes in Math.

Le problème que nous abordons ici est le suivant : soit (X_t) un mouvement brownien standard ; McKean nous a appris à considérer la mesure de Wiener comme la répartition uniforme de probabilité sur la ⟨⟨ sphère de Wiener ⟩⟩ de rayon unité, et lorsqu'on fait une dilatation de rapport σ on change de sphère et on obtient une mesure étrangère. Existe-t-il une manière naturelle de définir ⟨⟨ la même ⟩⟩ fonctionnelle sur des sphères de Wiener de rayons différents ? Autrement dit, F désignant une v.a. définie sur l'espace du mouvement brownien standard, peut on lui associer une v.a. F^σ sur l'espace du mouvement brownien de variance $\sigma^2 t$ de sorte que

-- aux constantes, aux applications coordonnées X_t, correspondent les mêmes constantes, les applications coordonnées X_t^σ de même indice ;
-- la correspondance soit linéaire et <u>multiplicative</u> ?

Nous donnerons une formule explicite pour F^σ, au moyen du développement de F suivant les chaos de Wiener, ainsi que des commentaires sur cette formule. Bien que le résultat soit très simple, il ouvre de nombreux problèmes intéressants. Nous avons du mal à croire qu'il ne figure nulle part dans l'abondante littérature sur les chaos de Wiener, mais nos recherches bibliographiques n'ont ramené à la surface qu'une formule de K.O. Friedrichs, qui traite un problème voisin, et que nous commenterons à la fin de cette note.

1. <u>Notations</u>. Nous travaillerons uniquement sur des fonctionnelles réelles F du mouvement brownien standard (X_t) , données par leur développement suivant les chaos de Wiener

$$(1) \qquad F = \Sigma_n \frac{1}{n!} \int_{\mathbb{R}_+^n} f_n(s_1,\ldots,s_n)dX_{s_1}\ldots dX_{s_n}$$

où la fonction f_n est <u>symétrique</u> et appartient à $L^2(\mathbb{R}_+^n)$ pour tout n .

On a $\quad \|F\|^2 = \Sigma_n \frac{1}{n!}\|f_n\|^2_{L^2(\mathbb{R}^n_+)}$.

Nous écrirons en abrégé, en englobant le coefficient 1/n! dans l'intégrale

(2) $\quad F = \Sigma_n \frac{1}{n!} I_n(f_n) = \Sigma_n J_n(f_n) = J(f_.)$

Il est parfois avantageux de permettre aux fonctions f_n de n'être pas symétriques, en convenant que $J_n(f_n)=J_n(Sf_n)$, la symétrisée de f : cela simplifie certains raisonnements combinatoires.

Par exemple, l'exponentielle stochastique $\mathcal{E}(u)$ ($u\varepsilon L^2(\mathbb{R}_+)$) admet une représentation (2) avec $f_n = u^{\otimes n} = u^{on}$ (o désigne le produit symétrique). Rappelons que $\mathcal{E}(u)=\exp(\int u_s dX_s - \frac{1}{2}\|u\|^2)$.

Ceci concerne le mouvement brownien standard ; pour le mouvement brownien de variance σ^2, il sera bon de faire apparaître explicitement le paramètre en écrivant $J^\sigma_n(f_n)$, $J^\sigma(f_.)$, $\mathcal{E}_\sigma(u)=\exp(\int u_s dX^\sigma_s - \sigma^2\|u\|^2/2)$

(et il pourra être bon aussi d'écrire J^1, etc. lorsque $\sigma=1$!). La formule donnant $\|F\|^2$ comporte, pour $\sigma\neq 1$, un facteur σ^{2n} devant le n-ième terme.

Nous aurons besoin aussi de la formule de multiplication des intégrales stochastiques, qui s'écrit ainsi, si $F=J^\sigma_m(f_m)$, $G=J^\sigma_n(g_n)$

(3) $$F \underset{\sigma}{\cdot} G = \Sigma_{p\leq m\wedge n} \frac{\sigma^{2p}}{p!} \frac{(m+n-2p)!}{(m-p)!(n-p)!} J_{m+n-2p}(f_m \underset{p}{\equiv} g_n)$$

où $f_m \underset{p}{\equiv} g_n$ désigne la fonction non symétrique obtenue en contractant, dans la fonction de m+n variables $f_m \otimes g_n$, p variables de f_m avec p variables de g_n. Cette formule se déduit de la formule de multiplication usuelle en l'exprimant au moyen des intégrales $J_n=I_n/n!$ (voir par ex. Sém. Prob. XX, LN 1204, p. 274). On peut l'exprimer autrement, en introduisant la fonction symétrique de m+n-2p variables

(4) $$\begin{aligned} & f_m \underset{p}{\approx} g_n \,(u_1,\ldots u_{m+n-2p}) = \\ & = \Sigma \int f_m(u_{i_1},\ldots,u_{i_{m-p}},r_1,\ldots,r_p) g_m(r_1,\ldots,r_p,u_{j_1}\ldots,u_{j_{n-p}}) dr_1\ldots dr_p \end{aligned}$$

la somme portant sur tous les choix $i_1<\ldots<i_{m-p}$ de m-p indices croissants parmi 1,...,m+n-2p (et le choix complémentaire étant $j_1<\ldots<j_{n-p}$ pour les indices restants). Alors on a simplement

(5) $$J_m(f_m) \underset{\sigma}{\cdot} J_n(g_n) = \Sigma_{p\leq m\wedge n} J_{m+n-2p}(f_m \underset{p}{\approx} g_n) \frac{\sigma^{2p}}{p!} .$$

Nous reviendrons plus loin sur cette formule.

2. La formule principale. Considérons l'exponentielle stochastique $\mathcal{E}(u)$ $=\exp(\int u_s dX_s - \|u\|^2/2$) ; d'après les règles que nous avons posées dans l'introduction, les v.a. $\exp(\int u_s dX^\sigma_s)$ sont «les mêmes» sur tous

les espaces de Wiener. Par conséquent, le prolongement de la v.a. $\mathcal{E}(u)$ à l'espace de Wiener de paramètre σ est

$$(\mathcal{E}(u))^{\sigma} = \exp(\frac{\sigma^2-1}{2}\|u\|^2)\mathcal{E}_{\sigma}(u)$$

Remplaçant u par λu et développant suivant les chaos, nous trouvons la formule

$$(6) \qquad (J_n(u^{\circ n}))^{\sigma} = \Sigma_{2k\leq n} \frac{1}{k!}(\frac{\sigma^2-1}{2})^k \|u\|^{2k} J^{\sigma}_{n-2k}(u^{\circ n-2k})$$

que nous allons interpréter, et généraliser.

Etant donnée une fonction symétrique f_n de n variables, suffisamment régulière (nous tenterons de préciser cela plus loin), nous considérons la fonction symétrique de n-2 variables, que nous appellerons la trace de f_n et noterons $\mathrm{Tr}f_n$

$$(7) \qquad \mathrm{Tr}f_n(s_1,\ldots,s_{n-2}) = \int f_n(s_1,\ldots,s_{n-2},s,s)ds \;.$$

On peut itérer l'opération, et définir $\mathrm{Tr}^k f_n$ pour $k\leq n/2$ (pour $k>n/2$ on conviendra que $\mathrm{Tr}^k f_n=0$). Par exemple, on a $\mathrm{Tr}^k(u^{\circ n})=\|u\|^{2k}u^{\circ n-2k}$ pour $k\leq n/2$. On peut donc écrire la formule (6) sous la forme

$$(J_n(u^{\circ n}))^{\sigma} = \Sigma_{2k\leq n} \frac{1}{k!}(\frac{\sigma^2-1}{2})^k J^{\sigma}_{n-2k}(\mathrm{Tr}^k u^{\circ n})$$

Désignons par $f_{\bullet}$ la famille des coefficients du développement de F suivant les chaos, à la manière de la formule (2), par $\mathrm{Tr}(f_{\bullet})$ la famille des coefficients $\mathrm{Tr}f_n$. On obtient alors la formule générale qui répond à la question de l'introduction

$$(8) \qquad F^{\sigma} = (J(f_{\bullet}))^{\sigma} = J^{\sigma}(e^{\frac{\sigma^2-1}{2}\mathrm{Tr}} f_{\bullet}) \;.$$

Jusqu'à maintenant nous avons prolongé de la sphère de Wiener de variance 1 à la sphère de variance σ^2 ; nous aurions aussi bien pu partir avec la variance initiale τ^2

$$(9) \qquad (J^{\tau}(f_{\bullet}))^{\sigma} = J^{\sigma}(e^{\frac{\sigma^2-\tau^2}{2}\mathrm{Tr}} f_{\bullet}) \;.$$

Cette formule met en évidence la compatibilité des prolongements : pour prolonger de la sphère τ à la sphère σ on peut aller de τ à ρ , puis de ρ à σ . Soulignons que la valeur 0 du paramètre n'est pas exclue : l'espace de Wiener de variance 0 est l'espace de Cameron-Martin, l'intégrale stochastique $I_n(f_n)$ étant la fonctionnelle

$$x \to \int f_n(s_1,\ldots,s_n)\dot{x}(s_1)\ldots\dot{x}(s_n)ds_1\ldots ds_n$$

et le produit (3) étant le produit de Wick.

3. <u>Justification de la formule</u>. La formule (8) ou (9) est linéaire, et elle a été vérifiée lorsque $f_n=u^{\circ n}$; par polarisation elle s'étend à une suite $(f_{\bullet})$ qui appartient à l'algèbre symétrique sur $L^2(\mathbb{R}_+)$ non complétée. A partir du cas des exponentielles, on vérifie sans peine

que l'application (9) transforme le produit $\underset{\tau}{\bullet}$ en le produit $\underset{\sigma}{\bullet}$. Il reste alors à compléter l'algèbre symétrique pour des ⟨⟨ normes-trace ⟩⟩ convenables, de manière à atteindre une classe naturelle de fonctionnelles auxquelles s'applique la formule (9).

Nous avons préféré donner à la formule (9) un sens différent. Pour éviter les difficultés liées à l'intégration sur toute la diagonale, nous nous plaçons sur un intervalle de temps [0,T] fini. Nous désignons alors par $\mathcal{F}$ l'ensemble des suites $f = (f_n)$ de fonctions symétriques, boréliennes bornées, nulles pour n suffisamment grand ; <u>soulignons qu'il s'agit ici de fonctions et non de classes</u>. Nous identifions la suite qui vaut 0 aux niveaux $m \neq n$ et f_n au niveau n à la fonction f_n, de sorte qu'il est est aussi légitime d'utiliser une notation additive $f = \Sigma_n f_n$. Nous définissons les produits $\underset{\sigma}{\bullet}$ sur $\mathcal{F}$ par

$$(10) \qquad f_m \underset{\sigma}{\bullet} g_n = \Sigma_{p \leq m\wedge n} \frac{\sigma^{2p}}{p!}(f_m \underset{p}{\approx} g_n) \quad (\text{ cf. } (5))$$

que l'on étend à $\mathcal{F}$ par linéarité. Nous écrivons la formule (9) comme une définition sur $\mathcal{F}$, en omettant les symboles d'intégrale stochastique

$$(11) \qquad U_{\sigma\tau} f = \Sigma_k \left(\frac{\sigma^2-\tau^2}{2}\right)^k/k! \quad Tr^k f$$

et alors le résultat, qui se démontre de façon purement combinatoire, mais sans mystère, est le suivant : les multiplications $\underset{\sigma}{\bullet}$ sont associatives, et $U_{\sigma\tau}$ transforme le produit $\underset{\tau}{\bullet}$ en produit $\underset{\sigma}{\bullet}$.

L'idée sous-jacente semble être que, pour travailler sur tous les espaces de Wiener à la fois, il faut se donner non seulement les coefficients du développement de F suivant les chaos (d'un point de vue physique, un élément de l'espace de Fock), mais aussi les ⟨⟨ coefficients fantômes ⟩⟩ sur les diagonales. Lorsque la fonctionnelle F est suffisamment régulière, la valeur de f_n sur les diagonales peut être déduite de la connaissance de la classe f_n : par exemple, pour n=2, si f_2 définit non seulement un opérateur de Hilbert-Schmidt mais un opérateur à trace, on peut donner un sens à la trace de f_2 sur la diagonale en développant $f(s,t)$ en $\Sigma_n a_{nm} e_n(s) e_m(t)$ dans une base o.n. et en faisant s=t dans cette formule ; cela devrait pouvoir s'étendre aux niveaux supérieurs, mais on sait que les opérateurs à trace sont rares, et il vaut sans doute mieux calculer sur les coefficients-fantômes !

4. <u>Quelques commentaires et problèmes ouverts</u>. Cet exposé ne contient presque pas de mathématiques, mais surtout des considérations heuristiques. Cependant, l'idée de ⟨⟨ la même v.a. ⟩⟩ sur les divers espaces de Wiener semble intéressante. En voici des exemples

1) Considérons une intégrale stochastique multiple $\int f_n(s_1,..,s_n) dX^\sigma_{s_1} .. dX^\sigma_{s_n}$. Cette intégrale peut se calculer de deux manières. Soit la manière

d'Ito, qui revient à exclure les diagonales, en décidant que l'intégrale multiple est n! fois l'intégrale sur le simplexe croissant ouvert. Soit encore, en incluant les diagonales, et en remplaçant $dX_s^{\sigma 2}$ par ds (et les puissances supérieures par 0). Le cas n=2 montre qu'il s'agit en fait d'intégrales multiples au sens de Stratonovitch. Désignons par $I_n^\sigma(f_n)$ la première intégrale, par $\hat{I}_n^\sigma(f_n)$ la seconde. Il est clair que

$$\Sigma_n \frac{1}{n!} I_n^\sigma(u^{on}) = \mathcal{E}_\sigma(u) \ , \quad \Sigma_n \frac{1}{n!} \hat{I}_n^\sigma(u^{on}) = \exp(\textstyle\int u_s dX_s^\sigma)$$

d'où l'on tire les conjectures suivantes (faciles à vérifier) : 1) les $\hat{I}_n^\sigma(f_n)$ sont ⟨⟨ la même v.a. ⟩⟩ sur les différents espace de Wiener ; 2) Il existe une formule simple permettant de passer de $\hat{I}_n^\sigma(f_n)$ à $I_n^\sigma(f_n)$ et vice-versa, et faisant intervenir les traces de f_n. Cette formule a été donnée, dans un contexte un peu différent, par Friedrichs (les $I_n^\sigma(f_n)$ sont les ⟨⟨ generalized Hermite polynomials ⟩⟩ , dans Mathematical aspects of the quantum theory of fields, p. 55 (Interscience, New York 1953).

2) Cette liaison entre le calcul de Stratonovitch et la formule (9) amène à se poser la question très naturelle suivante : est ce que les solutions d'une même équation différentielle stochastique de Stratonovitch, correspondant aux différentes valeurs du paramètre σ, sont ⟨⟨ la même ⟩⟩ v.a. en notre sens ? Cela amène à se poser des problèmes sur le cas déterministe (concernant l'existence des traces) qui ne font pas partie du folklore habituel.

3) Y.Z. Hu a résolu un problème analogue à celui que nous avons traité, mais dans une situation un peu différente : désignons par (X_t^x) le processus des accroissements du mouvement brownien standard issu de x. On se propose de définir l'<<égalité>> de deux fonctionnelles de deux mouvements browniens standard d'origines différentes (0 et x pour commencer). Il s'agit donc d'associer à une fonctionnelle F du mouvement brownien issu de 0 (donnée comme $F=J^0(f_.)$) une fonctionnelle du mouvement brownien issu de x (notée $F_x=J^x(f_.^x)$ de telle sorte que $1_x=1$, $(X_t)_x=x+X_t^x$, et que la correspondance soit multiplicative. Voici la formule obtenue par Hu

$$(J^y(f_.))_x = J^x(e^{(x-y)S}f_.)$$

où S est l'opérateur qui transforme la fonction symétrique $f_n(s_1,..,s_n)$ en $f_n(s_1,...,s_{n-1},0)$.

LIMIT THEOREM FOR ONE-DIMENSIONAL DIFFUSION PROCESS IN BROWNIAN ENVIRONMENT

Hiroshi TANAKA
Department of Mathematics. Faculty of Science and Technology
Keio University, Yokohama, 223 (Japan)

INTRODUCTION

Let $\mathbb{W}$ be the space of continuous functions $W: \mathbb{R} \to \mathbb{R}$ with $W(0) = 0$. In this paper an element of $\mathbb{W}$ is called an environment. Given an environment W, Brox[1] considered a diffusion process starting at 0 with generator

$$(1) \qquad \mathcal{L}_W = \frac{1}{2} e^{W(x)} \frac{d}{dx}\left(e^{-W(x)} \frac{d}{dx}\right) .$$

Such a diffusion, denoted by $X(t, W)$, is constructed from a one-dimensional Brownian motion $B(t)$ by a scale-change and a time-change. The probability measure governing $B(t)$ is denoted by P. We consider the Wiener measure Q on $\mathbb{W}$. Thus $\{W(x), x \geq 0, Q\}$ and $\{W(-x), x \geq 0, Q\}$ are independent one-dimensional Brownian motions starting at 0. We assume that $B(t)$ and $W(x)$ are independent so the full distribution governing $X(t, W)$ is $\mathcal{P} = P \otimes Q$. If $W(\cdot)$ were smooth, $X(t, W)$ would satisfy

$$X(t) = \text{a Brownian motion} - \frac{1}{2}\int_0^t W'(X(s))ds .$$

Although our $W(\cdot)$ is never smooth, the above remark will explain that $X(t, W)$ is regarded as a diffusion analogue of Sinai's random walk in a random environment([11]). $X(t, W)$ is called a diffusion process in a Brownian environment. The problem is to study the limiting behavior of $X(t, W)$ as $t \to \infty$. Brox[1] obtained the following result which is analogous to that of Sinai[11]: For any $\varepsilon > 0$

$$P\{(\log t)^{-2}X(t, W) \in U_\varepsilon(t, W)\} ,$$

which is regarded as a W-random variable, converges to 1 in probability as $t \to \infty$, where $U_\varepsilon(t, W)$ is the ε-neighborhood of $b_t(W)$ which is defined suitably in terms of "valleys" of the environment. The distribution of $b_t(W)$ is independent of t so the full distribution of $(\log t)^{-2}X(t, W)$ converges to that of $b_1(W)$. Kesten[7] obtained the exact form of the limit distribution. Kesten's result was then extended to the case of symmetric stable environments([12]).

The purpose of this paper is to elaborate Brox's result. We prove that, without scaling but only by centering, $X(t,\cdot)$ has a limit distribution as $t \to \infty$. To state the result more precisely, put $b(t, W) = (\log t)^2 b_t(W)$ and let $\tilde{Q}$ be the probability measure on $\mathbb{W}$ such that $\{W(x), x \geq 0, \tilde{Q}\}$ and $\{W(-x), x \geq 0, \tilde{Q}\}$ are independent Bessel processes of index 3 starting at 0. Note that $e^{-W} \in L^1(\mathbb{R})$ with $\tilde{Q}$-measure 1. Let Ω be the space of continuous paths $\omega:[0, \infty) \to \mathbb{R}$ and, for each W with $e^{-W} \in L^1(\mathbb{R})$, denote by $\tilde{P}_W$ the probability measure on Ω such that $\{\omega(t), t \geq 0, \tilde{P}_W\}$ is a diffusion process with generator (1) and with initial distribution

$$\tilde{\mu}_W(dx) = e^{-W(x)}dx \Big/ \int_{-\infty}^{\infty} e^{-W(y)}dy .$$

Finally put $\tilde{\mu} = \int \tilde{\mu}_W \tilde{Q}(dW)$ and $\tilde{P} = \int \tilde{P}_W \tilde{Q}(dW)$. Then our result is stated as follows.

Theorem. The process $\{X(t_0 + t, W) - b(t_0, W), t \geq 0, \mathcal{P}\}$ converges as $t_0 \to \infty$ to the stationary process $\{\omega(t), t \geq 0, \tilde{P}\}$ in the sense of weak convergence of probability measures on Ω. In particular the distribution of $X(t,\cdot) - b(t,\cdot)$ converges to $\tilde{\mu}$ as $t \to \infty$.

Similar results were also obtained by Golosov[2] for a reflecting random walk in random environment. Our method is on the extension line of Brox's and uses fine results on one-dimensional Brownian motion obtained by Lévy[8], Itô and McKean[4] and others.

§1. OUTLINE OF BROX'S METHOD

1.1. Let Ω_0 be the space of continuous paths $\omega:[0, \infty) \to \mathbb{R}$ with $\omega(0) = 0$ and let P be the Wiener measure on Ω_0. We write $B(t)$ for $\omega(t)$, the value of ω at time t. Thus $\{B(t), t \geq 0, P\}$ is a Brownian motion starting at 0. For a fixed $W \in \mathbb{W}$ we set

$$S(x) = \int_0^x e^{W(y)}dy ,$$

$$S^{-1}(y) = \text{the inverse function of } S(x),$$

$$A(s) = \int_0^s e^{-2W(S^{-1}(B(r)))}dr,$$

$$A^{-1}(t) = \text{the inverse function of } A(s).$$

Then $X(t, W) = S^{-1}(B(A^{-1}(t)))$ is a diffusion process with generator (1) starting at 0. If we set $(W^{x_0})(\cdot) = W(\cdot + x_0) - W(x_0)$, then

$X^{x_0}(t, W) = x_0 + X(t, W^{x_0})$ is a diffusion process with generator (1) starting at x_0.

Regarding W as a random element as well as ω, we have a process $X(t,\cdot)$ defined on the product probability space $\{\Omega \times \mathbb{W}, \mathcal{P} = P \otimes Q\}$. This full process is denoted by $\{X(t,\cdot)\}$ to distinguish it from the diffusion process $\{X(t, W), t \geq 0, P\}$ with a fixed W.

For $\lambda > 0$ and $W \in \mathbb{W}$ we define $W_\lambda \in \mathbb{W}$ by $W_\lambda(x) = \lambda^{-1} W(\lambda^2 x)$, $x \in \mathbb{R}$. The following scaling relation is important([1]): For any fixed $\lambda > 0$ and $W \in \mathbb{W}$

(1.1) $$\{X(t, \lambda W_\lambda), t \geq 0, P\} \overset{d}{=} \{\lambda^{-2} X(\lambda^4 t, W), t \geq 0, P\}$$

where $\overset{d}{=}$ means the equality in distribution.

1.2. We give the definition of a valley. Let $W \in \mathbb{W}$. A part $\{W(x), a \leq x \leq c\}$ of W is called a *valley* of W if

(i) $a < c$,

(ii) there exists $b \in (a, c)$ such that

$$W(a) > W(x) > W(b) \quad \text{for every } x \in (a, b) ,$$

$$W(c) > W(x) > W(b) \quad \text{for every } x \in (b, c) ,$$

(iii) for the same b as above

$$H_- \equiv \sup\{W(y) - W(x) : a \leq x < y \leq b\} < W(c) - W(b) ,$$

$$H_+ \equiv \sup\{W(x) - W(y) : b \leq x < y \leq c\} < W(a) - W(b) .$$

The value b in the above definition is particularly important and so, to stress b and also for simplicity, we write (a, b, c) instead of $\{W(x), a \leq x \leq c\}$. $A = H_+ \vee H_-$ is called the *inner directed ascent* (abbreviated to i.d.a.) and $D = (W(a) - W(b)) \wedge (W(c) - W(b))$ is the *depth* of the valley. When the letters A and D are used for notation they always mean the i.d.a. and the depth of (a,b,c). A valley (a,b,c) is said to contain 0 if $a < 0 < c$. It is easy to see that if (a, b, c) and (a', b', c') are valleys of W both containing 0 and satisfying $A < r < D$, $A' < r < D'$ (r is a positive constant), then $b = b'$. It is known that for any $r > 0$ and for any W in *some subset* $\tilde{\mathbb{W}}$ ($\subset \mathbb{W}$) with Q-measure 1 there exists a valley (a, b, c) of W with $A < r < D$ and containing 0 (see[1]). We denote by $b(W)$ the unique b of such a valley (a, b, c) for $r = 1$.

To give another description of $b(W)$ we put

$$W^{\#}(x) = \begin{cases} W(x) - \min\limits_{[0,x]} W & \text{for } x \geq 0, \\ W(x) - \min\limits_{[x,0]} W & \text{for } x < 0, \end{cases}$$

$$d_+ = \inf\{x > 0 : W^{\#}(x) = 1\},$$

$$d_- = \sup\{x < 0 : W^{\#}(x) = 1\},$$

$$V_+ = \min_{[0,d_+]} W, \quad V_- = \min_{[d_-,0]} W,$$

and define b_+ and b_- by $W(b_\pm) = V_\pm$ (such $b_\pm$ are uniquely determined with Q-measure 1). We also set

$$M_+ = \max_{[0,b_+]} W, \quad M_- = \max_{[b_-,0]} W.$$

Then another description of $b(W)$ is given as follows(see[7]):

$$(1.2) \qquad b(W) = \begin{cases} b_+ & \text{if } M_+ \vee (V_+ + 1) < M_- \vee (V_- + 1), \\ b_- & \text{if } M_+ \vee (V_+ + 1) > M_- \vee (V_- + 1). \end{cases}$$

Moreover, if we define $a(W)$ and $c(W)$ by

$a(W)$ = the infimum of the set of a's $(a < b(W))$ such that

$$\begin{cases} W(a) > W(x) > W(b(W)) & \text{for every } x \in (a, b(W)), \\ \sup\{W(y) - W(x) : a \leq x < y \leq b(W)\} < 1, \end{cases}$$

$c(W)$ = the supremun of the set of c's $(c > b(W))$ such that

$$\begin{cases} W(c) > W(x) > W(b(W)) & \text{for every } x \in (b(W), c), \\ \sup\{W(x) - W(y) : b(W) \leq x < y \leq c\} < 1. \end{cases}$$

Then $(a(W), b(W), c(W))$ is the maximum valley of W containing 0 and satisfying $A(W) < 1 < D(W)$, and is called the *standard valley* of W. Note that $a(W)$, $b(W)$, etc., are Borel functions on $\widetilde{W}$.

1.3. Let (a, b, c) be a valley of W and, for $\lambda > 0$, let T_λ^x be the exit time from (a, c) for the diffusion process $X^x(t, \lambda W)$. The following lemma is due to Brox[1].

Lemma 1 *([1]). For any* $\delta > 0$ *and a closed interval* $I \subset (a, c)$

$$(1.3) \qquad \lim_{\lambda\to\infty} \inf_{x\in I} P\{e^{\lambda(D - \delta)} < T_\lambda^x < e^{\lambda(D + \delta)}\} = 1 .$$

1.4. Brox([1]) employs a coupling technique. To explain it we had better adopt the path space representation of $X(t, \lambda W)$. So let $\Omega = C([0, \infty) \to \mathbb{R})$ and denote by $P_{\lambda W}$ the probability measure on Ω induced by the diffusion process $X(t, \lambda W)$. Moreover, we use the following notation. For an arbitrary interval $[a, c]$ and an environment W we denote by $P^\lambda_{W[a,c]}$ the probability measure on the path space $\Omega_{[a,c]} = C([0, \infty) \to [a, c])$ induced by the diffusion process on $[a, c]$ with (local) generator $\mathcal{L}_{\lambda W}$, with reflecting barriers at a and c and with initial distribution

$$\mu^{\lambda}_{W[a,c]}(dx) = e^{-\lambda W(x)}dx \Big/ \int_a^c e^{-\lambda W(y)}dy .$$

This reflecting diffusion is stationary since $\mu^{\lambda}_{W[a,c]}$ is its invariant measure. In particular, in case (a, b, c) is the standard valley of W, $P^{\lambda}_{W[a,c]}$, $\mu^{\lambda}_{W[a,c]}$ and $\Omega_{[a,c]}$ are abbreviated to P^{λ}_W, μ^{λ}_W and $\hat{\Omega}$, respectively.

We now assume that (a, b, c) is the standard valley of W. Let ω and $\hat{\omega}$ stand for generic elements of Ω and $\hat{\Omega}$ with values $\omega(t)$ and $\hat{\omega}(t)$ at time t, respectively, and consider two processes $\{\omega(t), t \geq 0\}$ and $\{\hat{\omega}(t), t \geq 0\}$ defined on the product probability space $\{\Omega^{\lambda} \times \hat{\Omega}, \mathbb{P}^{\lambda}_W\}$ where $\mathbb{P}^{\lambda}_W = P_{\lambda W} \otimes P^{\lambda}_W$. Thus the two processes are independent. Put

$$R = \inf\{t \geq 0 : \omega(t) = \hat{\omega}(t)\} ,$$
$$T_R = \inf\{t \geq R : \omega(t) \notin (a, c)\} ,$$
$$\hat{T}_R = \inf\{t \geq R : \hat{\omega}(t) \notin (a, c)\} .$$

Notice that these are random variables defined on $\{\Omega \times \hat{\Omega}, \mathbb{P}^{\lambda}_W\}$. If we define a process $\{\omega'(t), t \geq 0\}$ by

$$\omega'(t) = \begin{cases} \omega(t) & \text{for } 0 \leq t \leq R , \\ \hat{\omega}(t) & \text{for } t > R , \end{cases}$$

then

$$\{\omega(t), 0 \leq t \leq T_R, \mathbb{P}^{\lambda}_W\} \stackrel{d}{=} \{\omega'(t), 0 \leq t \leq \hat{T}_R, \mathbb{P}^{\lambda}_W\} . \tag{1.4}$$

The following lemma is also due to Brox[1]; the equality is a consequence of (1.4).

Lemma 2 *([1]). For any r_1 and r_2 such that $A < r_1 < r_2 < D$*

$$\mathbb{P}^{\lambda}_W\{R < e^{\lambda r_1} < e^{\lambda r_2} < T_R\} \tag{1.5}$$
$$= \mathbb{P}^{\lambda}_W\{R < e^{\lambda r_1} < e^{\lambda r_2} < \hat{T}_R\} \to 1, \quad \lambda \to \infty .$$

Using Lemma 2 and the scaling relation (1.1), Brox([1]) obtained his main results: For any $\varepsilon > 0$

$$P\{|\lambda^{-2}X(e^{\lambda}, W) - b(W_{\lambda})| > \varepsilon\} \to 0 \tag{1.6}$$

in probability with respect to Q as $\lambda \to \infty$.

By the same argument as Brox's we can obtain a refinement of his result as will be discussed in the next subsection.

1.5. We keep the notation of 1.4 and, in addition, we denote by θ_t (resp. $\hat{\theta}_t$) the shift on Ω (resp. $\hat{\Omega}$) defined by $(\theta_t\omega)(\cdot) = \omega(t+\cdot)$ (resp. $(\hat{\theta}_t\hat{\omega})(\cdot) = \hat{\omega}(t+\cdot)$). For $\lambda > 0$, γ_λ denotes the map: $\Omega \to \Omega$ defined by $(\gamma_\lambda\omega)(t) = \lambda^2\omega(\lambda^{-4}t)$, $t \geq 0$. $\mathcal{B}_t$ denotes the σ-field on Ω generated by the sets of the form $\{\omega: \omega(s) \leq x\}$, $0 \leq s \leq t$, $x \in \mathbb{R}$, and $\mathcal{B} = \bigvee \mathcal{B}_t$. For $\omega \in \Omega$ and $x \in \mathbb{R}$, $\omega - x$ denotes the path whose value at time t is $\omega(t) - x$; $\hat{\omega} - x$ (for $\hat{\omega} \in \Omega$) also denotes a similar path. The following notational convention is used:

(1.7a) $\quad P^\lambda_{W[a,c]}\{\Gamma\} = P^\lambda_{W[a,c]}\{\Gamma \cap \Omega_{[a,c]}\}, \quad \Gamma \in \mathcal{B},$

(1.7b) $\quad P^\lambda_W\{\hat{\omega} \in \Gamma\} = P^\lambda_W\{\Gamma \cap \hat{\Omega}\}, \quad \Gamma \in \mathcal{B}.$

Note that the right hand sides of the above make sense since both $\Omega_{[a,c]}$ and $\hat{\Omega}$ are measurable subsets of Ω.

For any family $\{r(\lambda)\}$ such that $r(\lambda) \to 1$ $(\lambda \to \infty)$ Lemma 2 implies

(1.8) $$\varepsilon_\lambda(W) \equiv 1 - \mathbb{P}^\lambda_W\{R < s(\lambda) < s(\lambda) + t(\lambda) < T_R\} \to 0$$

as $\lambda \to \infty$ for any $W \in \tilde{W}$ and the same is true with T_R replaced by $\hat{T}_R$, where

(1.9) $$s(\lambda) = \lambda^{-4}e^{\lambda}, \quad t(\lambda) = \lambda^{-4}e^{\lambda r(\lambda)}.$$

We are now in position to state

Refinement of Brox's result. *For $\{r(\lambda)\}$ as above and for any $\Gamma_\lambda \in \mathcal{B}_{u(\lambda)}$, $u(\lambda) = e^{\lambda r(\lambda)}$, $\lambda > 0$, we have*

(1.10) $$P_W\{\theta_{exp\lambda}\omega - \lambda^2 b(W_\lambda) \in \Gamma_\lambda\}$$
$$\stackrel{d}{=} P^\lambda_W\{\hat{\omega} - b(W) \in \gamma_\lambda^{-1}(\Gamma_\lambda)\}^{*)} + \varepsilon_\lambda(W, \Gamma_\lambda),$$

where $b(\cdot)$ is defined by (1.2) and $\varepsilon_\lambda(\cdot, \Gamma_\lambda)$ is a suitable random variable defined on $(\tilde{W}, Q)$ satisfying

(1.11) $$|\varepsilon_\lambda(W, \Gamma_\lambda)| \leq \varepsilon_\lambda(W).$$

The proof is as follows. Since the scaling relation (1.1) implies

---

*) The convention (1.7b) is used.

$$\{(\theta_{\exp\lambda}\omega)(t),\ t \geq 0,\ P_W\} \overset{d}{=} \{(\gamma_\lambda \theta_{s(\lambda)}\omega)(t),\ t \geq 0,\ P_{\lambda W_\lambda}\},$$

using the notation (1.9) we have

$$\begin{aligned}
&P_W\{\theta_{\exp\lambda}\omega - \lambda^2 b(W_\lambda) \in \Gamma_\lambda\} \\
&= P_{\lambda W_\lambda}\{\gamma_\lambda \theta_{s(\lambda)}\omega - \lambda^2 b(W_\lambda) \in \Gamma_\lambda\} \\
&\overset{d}{=} P_{\lambda W}\{\gamma_\lambda \theta_{s(\lambda)}\omega - \lambda^2 b(W) \in \Gamma_\lambda\} \qquad (\text{since } W_\lambda \overset{d}{=} W) \\
&= \mathbb{P}_W^\lambda\{R < s(\lambda),\ \theta_{s(\lambda)}\omega - b(W) \in \gamma_\lambda^{-1}(\Gamma_\lambda),\ s(\lambda)+t(\lambda) < T_R\} + \varepsilon_\lambda^0 \\
&\qquad\qquad (\text{by } (1.8)) \\
&= \mathbb{P}_W^\lambda\{R < s(\lambda),\ \hat{\theta}_{s(\lambda)}\omega - b(W) \in \gamma_\lambda^{-1}(\Gamma_\lambda),\ s(\lambda)+t(\lambda) < \hat{T}_R\} + \varepsilon_\lambda^0 \\
&\qquad\qquad (\text{by } (1.4)) \\
&= P_W^\lambda\{\hat{\theta}_{s(\lambda)}\hat{\omega} - b(W) \in \gamma_\lambda^{-1}(\Gamma_\lambda)\} + \varepsilon_\lambda \qquad (\text{by } (1.8) \text{ for } \hat{T}_R) \\
&= P_W^\lambda\{\hat{\omega} - b(W) \in \gamma_\lambda^{-1}(\Gamma_\lambda)\} + \varepsilon_\lambda,
\end{aligned}$$

where $\varepsilon_\lambda^0 = \varepsilon_\lambda^0(W, \Gamma_\lambda)$ and $\varepsilon_\lambda = \varepsilon_\lambda(W, \Gamma_\lambda)$ are suitable random variables satisfying (1.11).

Before ending this section we state one more lemma.

<u>Lemma 3</u>. *Let (a, b, c) be the standard valley of W $(\in \tilde{\mathbb{W}})$. If $\rho(\lambda)$, $\lambda > 0$, satisfies*

$$(1.12) \qquad \rho(\lambda) \geq 0 \ \text{ and } \ \rho(\lambda) = o(e^{\varepsilon\lambda}) \ \text{ as } \ \lambda \to \infty \ \text{ for } \forall\varepsilon > 0,$$

then for any Γ_λ, $\lambda > 0$, satisfying

$$(1.13) \qquad \Gamma_\lambda \in \mathcal{B}_{\rho(\lambda)}, \quad \lambda > 0,$$

and for any a' and c' with $a < a' < b < c' < c$ we have

$$(1.14) \qquad P_W^\lambda\{\hat{\omega} - b \in \Gamma_\lambda\} = P^\lambda_{W^b[a'-b,c'-b]}\{\Gamma_\lambda\} + o(1), \quad \lambda \to \infty,$$

where $o(1)$ is uniform with respect to the choice of Γ_λ under the condition (1.13) and $W^b(\cdot) = W(\cdot + b) - W(b)$.

Proof. Since for any $\delta > 0$ both $\mu_W^\lambda\{(b - \delta, b - \delta)\}$ and $\mu^\lambda_{W^b[a'-b,c'-b]}\{(-\delta,-\delta)\}$ tend to 1 as $\lambda \to \infty$, it follows from Lemma 1 and (1.12) that

$$(1.15) \qquad P_W^\lambda\{\hat{T} > \rho(\lambda)\} \to 1, \quad P^\lambda_{W^b[a'-b,c'-b]}\{T' > \rho(\lambda)\} \to 1 \quad (\lambda \to \infty),$$

where $\hat{T}$ and T' are the exit times of (a', c') and $(a'-b, c'-b)$, respectively, for the processes under consideration. Moreover, we see that

(1.16a) $\mu_W^\lambda(dx) = \kappa(\lambda)\mu^\lambda_{W[a',c']}(dx)$ on $[a', c']$,

(1.16b) $\kappa(\lambda) = \int_{a'}^{c'} e^{-\lambda W(x)}dx \Big/ \int_a^c e^{-\lambda W(x)}dx \to 1, \quad \lambda \to \infty$.

Therefore, we have as $\lambda \to \infty$

$$\begin{aligned}
&P_W^\lambda\{\hat{\omega} - b \in \Gamma_\lambda\}\\
&= P_W^\lambda\{\hat{\omega} - b \in \Gamma_\lambda, \hat{T} > \rho(\lambda)\} + o(1) \qquad \text{(by (1.15))}\\
&= \kappa(\lambda)P^\lambda_{W^b[a'-b,c'-b]}[\Gamma_\lambda \cap \{T' > \rho(\lambda)\}] + o(1) \qquad \text{(by 1.16a))}\\
&= P^\lambda_{W^b[a'-b,c'-b]}[\Gamma_\lambda \cap \{T' > \rho(\lambda)\}] + o(1) \qquad \text{(by 1.16b))}\\
&= P^\lambda_{W^b[a'-b,c'-b]}\{\Gamma_\lambda\} + o(1) \qquad \text{(by (1.15))} ,
\end{aligned}$$

completing the proof of the lemma.

§2. THE LAW OF THE STANDARD VALLEY.

Recalling the notation of 1.2 we put

$$W_+ = \{W(b_+ + t) - W(b_+), \ -b_+ \leq t \leq d_+ - b_+, \ Q\} ,$$

$$W_- = \{W(b_- - t) - W(b_-), \ b_- \leq t \leq -(d_- - b_-), \ Q\} .$$

On a suitable probability space we consider a process $\{x_+(t), t \in \mathbb{R}\}$ such that $\{x_+(t), t \geq 0\}$ and $\{x_+(-t), t \geq 0\}$ are independent reflecting Brownian motions on $[0, \infty)$ starting at 0 (abbreviation: RBM^0). Let $\{\ell_+(t), t \in \mathbb{R}\}$ be the local timeat 0 of $x_+(t)$, that is,

$$\ell_+(t) = \lim_{\varepsilon \downarrow 0} \frac{1}{2\varepsilon} \int_I \mathbb{1}_{[0,\varepsilon]}(x_+(s))ds ,$$

where $I = [0, t]$ or $[t, 0]$ according as $t \geq 0$ or $t < 0$. Also put $\beta_+(t) = x_+(t) + \ell_+(t)$. Then by Pitman's theorem([9]) $\{\beta_+(t), t \geq 0\}$ is a Bessel process of index 3 starting at 0 (abbreviation: $BES^0(3)$). We put

(2.1a) σ_+ = the smallest zero of $x_+(t)$ in $(z, 0]$ where z is the maximum of $t < 0$ with $x_+(t) = 1$,

(2.1b) $\tau_+ = \min\{t > 0 : \beta_+(t) = 1\}$.

Proposition. W_- *and* W_+ *are independent and*

$$W_- \overset{d}{=} W_+ \overset{d}{=} \{\beta_+(t),\ \sigma_+ \le t \le \tau_+\} . \tag{2.2}$$

Proof. The independence of W_- and W_+ and the first law equality in (2.2) are obvious. For the proof of the second law equality it is convenient to use the construction of an equivalent of W_+ by means of the excursions of a RBM^0 ([4][5][3]). Denote by $\mathcal{W}^+$ the space of $w:[0,\infty) \to \mathbb{R}^+$ satisfying

(i) $w(t) > 0$ for $0 < t < \zeta(w) = \min\{s > 0 : w(s) = 0\}$,

(ii) $w(0) = w(t) = 0$ for $t \ge \zeta(w)$.

We consider a σ-finite measure n^+ on $\mathcal{W}^+$ defined by

$$n^+\left[\{w(t_1) \in A_1,\ w(t_2) \in A_2, \cdots,\ w(t_n) \in A_n\}\right]$$

$$= \int_{A_1} K^+(t_1, x_1)dx_1 \int_{A_2} p^0(t_2-t_1,x_1,x_2)dx_2 \int_{A_3} \cdots$$

$$\cdots \int_{A_n} p^0(t_n-t_{n-1},x_{n-1},x_n)dx_n$$

where $0 < t_1 < t_2 < \cdots < t_n$, $A_i \in \mathcal{B}(\mathbb{R}^+)$, $1 \le i \le n$, and

$$K^+(t, x) = \sqrt{\frac{2}{\pi t^3}}\, e^{-x^2/2t} , \quad t > 0, \quad x \in \mathbb{R}^+,$$

$$p^0(t,x,y) = \frac{1}{\sqrt{2\pi t}}\{e^{-(x-y)^2/2t} - e^{-(x+y)^2/2t}\} ,$$

$$t > 0, \quad x, y \in \mathbb{R}^+.$$

Let $p(t)$ be a stationary Poisson point process on $\mathcal{W}^+$ with characteristic measure n^+ and set

$$a(t) = \sum_{0<s\le t} \zeta(p(s)) ,$$

$$\ell(t) = \text{the inverse function of } a(\cdot) .$$

We define $x(t)$ by

$$x(t) = \begin{cases} p(s)(t - a(s-)) & \text{if } a(s-) \le t \le a(s) \\ 0 & \text{if } a(s-) = t = a(s) , \end{cases}$$

where $s = \ell(t)$. Then $x(t)$ is a RBM^0 and $\ell(t)$ is its local time at 0 ([3]). For $w \in \mathcal{W}^+$ we denote by $h(w)$ the maximum value of $w(t)$ and put

$$\xi = \min\{s > 0 : h(p(s)) > 1\} .$$

$$\tau = \min\{t > 0 : p(\xi)(t) = 1\} .$$

Note that $\xi < \infty$ a.s. because $n^+[\{h(w) > 1\}] = 1 < \infty$. We also put

$$W_+^- = \{W(b_+ + t) - W(b_+),\ - b_+ \leq t \leq 0,\ Q\}\ ,$$

$$W_+^+ = \{W(b_+ + t) - W(b_+),\ 0 \leq t \leq d_+ - b_+,\ Q\}\ ,$$

$$W_+^0 = \{W(t),\ 0 \leq t \leq b_+,\ Q\}\ .$$

Since $\{W^\#(t),\ t \geq 0,\ Q\}$ is a RBM^0 and $-\min\{W(s):0\leq s\leq t\}$ is its local time at 0, we see that the joint distribution of W_+^0 and W_+^+ is the same as the joint distribution of the following processes (2.3) and (2.4):

(2.3) $\{x(t) - \ell(t),\ 0 \leq t \leq a(\xi-)\}$,

(2.4) $\{p(\xi)(t),\ 0 \leq t \leq \tau\}$.

On the other hand, $\{p(s),\ 0 \leq s < \xi\}$ and $\{p(\xi)(t),\ 0 \leq t \leq \tau\}$ are independent and hence the processes (2.3) and (2.4) are also independent. Therefore, W_+^+ is independent of W_+^0 and consequently of W_+^-. Moreover, since $\{p(\xi)(t),\ 0 \leq t \leq \tau\}$is a part of a $BES^0(3)$ ([13], see also [10]), it follows that

$$W_+^+ \overset{d}{=} \{\beta_+(t),\ 0 \leq t \leq \tau_+\}\ .$$

It remains to prove

(2.5) $W_+^- \overset{d}{=} \{\beta_+(t),\ \sigma_+ \leq t \leq 0\}$.

To prove this we must consider the time reversal of (2.3), or more precisely, of

(2.6) $\{x(t) - \ell(t) + \xi,\ \ 0 \leq t \leq a(\xi-)\}$.

Put

$$\mathcal{W}_0^+ = \{w \in \mathcal{W}^+:\ h(w) < 1\}\ ,$$

$$\mathcal{W}_1^+ = \{w \in \mathcal{W}^+:\ h(w) \geq 1\}\ ,$$

and define two point processes p_0 and p_1 by

$$p_i(t) = p(t) \qquad \text{for } t \in D_{p_i},\quad i = 0, 1\ ,$$

where D_{p_i} (the domain of definition of p_i) is given by

$$D_{p_i} = \{t \in (0, \infty):\ p(t) \in \mathcal{W}_i^+\},\quad i = 0, 1\ .$$

Then p_0 and p_1 are independent stationary Poisson point processes on $\mathcal{W}_0^+$ and $\mathcal{W}_1^+$ with characteristic measures $n_i^+ =$ the restriction of n^+ to $\mathcal{W}_i^+$, $i = 0, 1$, respectively. For $w \in \mathcal{W}_0^+$ we define $\hat{w} \in \mathcal{W}_0^+$ by

$$\hat{w}(s) = \begin{cases} w(\zeta(w) - s) & \text{for } 0 \leq s \leq \zeta(w) \\ 0 & \text{for } s > \zeta(w) . \end{cases}$$

Then the measure n_0^+ is invariant under the map: $w \to \hat{w}$. We define $\tilde{p}_0$ by

$$\tilde{p}_0(t) = \begin{cases} \widehat{p_0(\xi - t)} & \text{for } t < \xi \\ p_0(t) & \text{for } t \geq \xi . \end{cases}$$

It can be proved that $\tilde{p}_0$ is again a stationary Poisson point process on $\mathcal{W}_0^+$ with characteristic measure n_0^+ and is independent of p_1. Therefore, the point process $\tilde{p}$ defined by

$$\tilde{p}(t) = \begin{cases} \tilde{p}_0(t) & \text{for } t \in D_{\tilde{p}_0} \\ p_1(t) & \text{for } t \in D_{p_1} \end{cases}$$

is equivalent in law to p. Therefore, if we set

$$\tilde{a}(t) = \sum_{0<s\leq t} \zeta(\tilde{p}(t)) ,$$

$$\tilde{\ell}(t) = \text{the inverse function of } \tilde{a}(\cdot) ,$$

and if we define $\tilde{x}(t)$ by

$$\tilde{x}(t) = \begin{cases} \tilde{p}(s)(t - \tilde{a}(s-)) & \text{if } \tilde{a}(s-) \leq t \leq \tilde{a}(s) \\ 0 & \text{if } \tilde{a}(s-) = t = \tilde{a}(s) \end{cases}$$

where $s = \tilde{\ell}(t)$, then $\tilde{x}(t)$ is a RBM^0 and $\tilde{\ell}(t)$ is its local time at 0. Also we have $\tilde{a}(\xi-) = a(\xi-)$. Moreover, it is seen that (drawing a picture is helpful)

$$\tilde{x}(t) = x(a(\xi-) - t) , \qquad 0 \leq t \leq a(\xi-) ,$$

$$\begin{aligned} \tilde{\ell}(t) &= \ell(a(\xi-)) - \ell(a(\xi-) - t) \\ &= \xi - \ell(a(\xi-) - t) , \qquad 0 \leq t \leq a(\xi-) . \end{aligned}$$

Therefore, the time reversal of (2.6) is

$$\text{(2.7)} \qquad \begin{aligned} &\{x(a(\xi-) - t) - \ell(a(\xi-) - t) + \xi, \quad 0 \leq t \leq a(\xi-)\} \\ &\overset{d}{=} \{\tilde{x}(t) + \tilde{\ell}(t), \quad 0 \leq t \leq \tilde{a}(\xi-)\} . \end{aligned}$$

Since W_+^- is equivalent in law to

$$\{x(a(\xi-) + t) - \ell(a(\xi-) + t) + \xi, \quad -a(\xi-) \leq t \leq 0\} ,$$

(2.7) implies (2.5) as was to be proved.

§3. PROOF OF THE THEOREM

We are going to prove the theorem announced in the introduction with

(3.1) $$b(t,W) = (\log t)^2 b(W_{\log t}), \quad t > 1 .$$

Let $\mathbb{P}$ be the probability measure on $\Omega \times \mathbb{W}$ defined by $\mathbb{P}(d\omega dW) = P_W(d\omega)Q(dW)$. Then the theorem is rephased as follows: The process

$$\{\omega(e^\lambda + t) - \lambda^2 b(W_\lambda), \quad t \geq 0, \quad \mathbb{P}\}$$

converges in law to the process $\{\omega(t), t \geq 0, \tilde{P}\}$ as $\lambda \to \infty$.

In addition to the process $x_+(t)$ of §2, we need another process $\{x_-(t), t \in \mathbb{R}\}$ which is equivalent in law to $\{x_+(t), t \in \mathbb{R}\}$. We assume that $x_+(t)$ and $x_-(t)$ are defined on a common probability space $\{\bar{\Omega}, \bar{P}\}$ and that they are independent. Denote by $\ell_-(t)$ the local time at 0 of $x_-(t)$, put $\beta_-(t) = x_-(t) + \ell_-(t)$ and define σ_- and τ_- in a way similar to (2.1). The expectation with respect to $\bar{P}$ is denote by $\bar{E}$.

Let $\rho(\lambda)$, $\lambda > 0$, be a given function satisfying the condition (1.12) and let $\Gamma_\lambda \in \mathcal{B}_{\rho(\lambda)}$, $\lambda > 0$. Then the condition for Γ_λ in the refinement of Brox's result is automatically satisfied. Therefore, by (1.10) and (1.2) we have

(3.2) $$\begin{aligned} &E_Q[P_W\{\theta_{\exp\lambda}\omega - \lambda^2 b(W_\lambda) \in \Gamma_\lambda\}] \\ &= E_Q[P_W^\lambda\{\hat{\omega} - b(W) \in \gamma_\lambda^{-1}(\Gamma_\lambda)\}] + o(1) \\ &= I_\lambda + II_\lambda + o(1) , \end{aligned}$$

where

$$I_\lambda = E_Q[P_W^\lambda\{\hat{\omega} - b_+ \in \gamma_\lambda^{-1}(\Gamma_\lambda)\}; \; b = b_+]$$

$$II_\lambda = E_Q[P_W^\lambda\{\hat{\omega} - b_- \in \gamma_\lambda^{-1}(\Gamma_\lambda)\}; \; b = b_-] ,$$

and E_Q denotes the expectation with respect to Q while $E_Q\{\text{—}; b = b_+\}$ denotes the integral over the set $\{b = b_+\}$. It is to be noted that in the above (and also in what follows) $o(1)$ means a term which tends to 0 as $\lambda \to \infty$ uniformly with respect to the choice of $\{\Gamma_\lambda\}$ so far as it satisfies the condition (1.13). We are going to use (2.2) to compute I_λ and II_λ.

We put

$$\bar{b}_+ = -\sigma_+, \quad \bar{b}_- = \sigma_- ,$$

$$\bar d_+ = \tau_+ - \sigma_+, \quad \bar d_- = -(\tau_- - \sigma_-),$$
$$\bar M_+ = \max_{[\sigma_+,0]} \beta_+ - \beta_+(\sigma_+), \quad \bar M_- = \max_{[\sigma_-,0]} \beta_- - \beta_-(\sigma_-),$$
$$\bar V_+ = -\beta_+(\sigma_+), \quad \bar V_- = -\beta_-(\sigma_-),$$
$$\beta(t) = \begin{cases} \beta_+(\sigma_+ + t) - \beta_+(\sigma_+) & \text{for } t \geq 0, \\ \beta_-(\sigma_- - t) - \beta_-(\sigma_-) & \text{for } t < 0, \end{cases}$$

Then the proposition of §2 implies

$$(3.3) \qquad \{W(t),\ d_- \leq t \leq d_+\} \overset{d}{=} \{\beta(t),\ \bar d_- \leq t \leq \bar d_+\}.$$

As in (1.2) we define $\bar b$ by

$$\bar b = \begin{cases} \bar b_+ & \text{if } \bar M_+ \vee (\bar V_+ + 1) < \bar M_- \vee (\bar V_- + 1), \\ \bar b_- & \text{if } \bar M_+ \vee (\bar V_+ + 1) > \bar M_- \vee (\bar V_- + 1). \end{cases}$$

Using Lemma 3 and then (3.3) we have

$$(3.4) \qquad I_\lambda = E_Q[P^\lambda_{W^{b_+}[-b_+,\ d_+-b_+]}\{\gamma_\lambda^{-1}(\Gamma_\lambda)\};\ b = b_+] + o(1)$$
$$= \bar E[P^\lambda_{\beta_+[\sigma_+,\ \tau_+]}\{\gamma_\lambda^{-1}(\Gamma_\lambda)\};\ b = b_+] + o(1),$$

where $P^\lambda_{\beta_+[\cdot,\cdot]}$ is the probability measure defined in a way similar to $P^\lambda_{W[\cdot,\cdot]}$.

For $\varepsilon > 0$ put

$$\sigma_\varepsilon = \max\{t < 0 : x_+(t) = 0 \text{ and } t < \exists s < 0 \text{ s.t. } x_+(s) = \varepsilon\},$$
$$P^\lambda_{\beta_+,\varepsilon} = P^\lambda_{\beta_+[\sigma_\varepsilon,\tau_+]}.$$

Also let $\tilde\mu_{\lambda\beta_+}$ be the probability measure on $\mathbb{R}$:

$$\tilde\mu_{\lambda\beta_+}(dx) = e^{-\lambda\beta_+(x)}dx \Big/ \int_{-\infty}^{\infty} e^{-\lambda\beta_+(t)}dt,$$

and let $\tilde P_{\lambda\beta_+}$ be the probability measure on Ω defined in a way similar to $\tilde P_{\lambda W}$ (see the introduction). Then the scaling relation

$$(3.5) \qquad \{\lambda^{-1}\beta_+(\lambda^2 t),\ t \in \mathbb{R}\} \overset{d}{=} \{\beta_+(t),\ t \in \mathbb{R}\}$$

implies that for any $\delta > 0$

$$\tilde\mu_{\lambda\beta_+}\{(-\delta,\delta)\} \overset{d}{=} \int_{-\lambda^2\delta}^{\lambda^2\delta} e^{-\beta_+(x)}dx \Big/ \int_{-\infty}^{\infty} e^{-\beta_+(t)}dt \to 1, \quad \lambda \to \infty,$$

and hence $E[\tilde{\mu}_{\lambda\beta_+}\{(-\delta,\delta)\}] \to 1$. Similarly $E[\mu^{\lambda}_{\beta_+[\sigma_+,\tau_+]}\{(-\delta,\delta)\}] \to 1$ and $E[\mu^{\lambda}_{\beta_+[\sigma_\varepsilon,\tau_+]}\{(-\delta,\delta)\}] \to 1$ as $\lambda \to \infty$. Therefore, just as in the case of Lemma 3, we can prove the following: Let Γ_λ, $\lambda > 0$, be the same as before. Then

$$(3.6) \qquad \bar{E}[\tilde{P}_{\lambda\beta_+}\{\gamma_\lambda^{-1}(\Gamma_\lambda)\}] = \bar{E}[P^{\lambda}_{\beta_+[\sigma_+,\tau_+]}\{\gamma_\lambda^{-1}(\Gamma_\lambda)\}] + o(1)$$
$$= \bar{E}[P^{\lambda}_{\beta_+,\varepsilon}\{\gamma_\lambda^{-1}(\Gamma_\lambda)\}] + o(1), \quad \lambda \to \infty .$$

From (3.4) and (3.6) we have for any $\varepsilon > 0$

$$(3.7) \qquad I_\lambda = \bar{E}[P^{\lambda}_{\beta_+,\varepsilon}\{\gamma_\lambda^{-1}(\Gamma_\lambda)\};\ \bar{b} = \bar{b}_+] + o(1), \quad \lambda \to \infty ,$$

and a similar formula for II_λ.

We put

$$\tilde{x}_+(t) = x_+(\sigma_\varepsilon + t) - x_+(\sigma_\varepsilon) ,$$
$$\tilde{\beta}_+(t) = \beta_+(\sigma_\varepsilon + t) - \beta_+(\sigma_\varepsilon) ,$$

$\tilde{\sigma}_+$ = the smallest zero of $\tilde{x}_+(t)$ in $(z, 0]$ where z is the maximum of $t < 0$ such that $\tilde{x}_+(t) = 1$,

$$\tilde{M}_+ = \max_{[\tilde{\sigma}_+,0]} \tilde{\beta}_+ - \tilde{\beta}_+(\tilde{\sigma}_+) ,$$

$$m_+ = \max_{[\sigma_+,0]} \beta_+, \qquad m_\varepsilon = \max_{[\sigma_\varepsilon,0]} \beta_+ .$$

Then $m_\varepsilon < m_+$ for all sufficiently small $\varepsilon > 0$ ($\bar{P}$-a.s.) and

$$m_\varepsilon < m_+ \Longrightarrow \sigma_+ < \sigma_\varepsilon$$
$$\Longrightarrow \beta_+(\sigma_+) = \tilde{\beta}_+(\tilde{\sigma}_+) + \beta_+(\sigma_\varepsilon), \quad \bar{M}_+ = \tilde{M}_+ .$$

Therefore

$$\{\bar{b} = \bar{b}_+\} \cap \{m_\varepsilon < m_+\}$$
$$= \{\tilde{M}_+ \vee (1 - \tilde{\beta}_+(\tilde{\sigma}_+) - \beta_+(\sigma_\varepsilon)) < \bar{M}_- \vee (1 - \beta_-(\sigma_-))\} \cap \{m_\varepsilon < m_+\} ,$$

and hence

$$(3.8) \qquad \bar{E}[P^{\lambda}_{\beta_+,\varepsilon}\{\gamma_\lambda^{-1}(\Gamma_\lambda)\};\ \bar{b} = \bar{b}_+]$$
$$= \bar{E}[P^{\lambda}_{\beta_+,\varepsilon}\{\gamma_\lambda^{-1}(\Gamma_\lambda)\};\ \tilde{M}_+ \vee (1-\tilde{\beta}_+(\tilde{\sigma}_+)-\beta_+(\sigma_\varepsilon)) < \bar{M}_- \vee (1-\beta_-(\sigma_-))]$$
$$+ \Delta(\varepsilon,\lambda) ,$$

where $\Delta(\varepsilon,\lambda)$ tends to 0 as $\varepsilon\downarrow 0$ uniformly in λ. Since $P^{\lambda}_{\beta_+,\varepsilon}\{\gamma_\lambda^{-1}(\Gamma_\lambda)\}$ is a measurable function of $\{x_+(t),\ \sigma_\varepsilon\le t\le\tau_+\}$ and since the process $\{\tilde\beta_+(t),\ \tilde\sigma_+\le t\le 0\}$ conditioned by $\{x_+(t),\ \sigma_\varepsilon\le t\le\tau_+\}$ is equivalent in law to $\{\beta_+(t),\ \sigma_+\le t\le 0\}$, by using the strong Markov property of $x_+(t)$ we see that the right hand side of (3.8) equals

$$(3.9)\qquad \bar E[P^{\lambda}_{\beta_+,\varepsilon}\{\gamma_\lambda^{-1}(\Gamma_\lambda)\}\Psi(\beta_+(\sigma_\varepsilon))] + \Delta(\varepsilon,\lambda)\ ,$$

where $\psi(x) = \bar P\{\bar M_+^{\vee}(1-\beta_+(\sigma_+)-x) < \bar M_-^{\vee}(1-\beta_-(\sigma_-))\}$. Since $\beta_+(\sigma_\varepsilon)\to 0$ as $\varepsilon\downarrow 0$ and $\psi(x)\to 1/2$ as $x\to 0$, (3.9) is equal to

$$\frac{1}{2}\,\bar E[P^{\lambda}_{\beta_+,\varepsilon}\{\gamma_\lambda^{-1}(\Gamma_\lambda)\}] + \Delta'(\varepsilon,\lambda)$$

which, by (3.6), is again equal to

$$(3.10)\qquad \frac{1}{2}\,\bar E[\tilde P_{\lambda\beta_+}\{\gamma_\lambda^{-1}(\Gamma_\lambda)\}] + \Delta'(\varepsilon,\lambda) + \Delta''(\varepsilon,\lambda)\ ,$$

where $\Delta'(\varepsilon,\lambda)\to 0$ as $\varepsilon\downarrow 0$ uniformly in λ and $\Delta''(\varepsilon,\lambda)\to 0$ as $\lambda\to\infty$ for each fixed $\varepsilon>0$. Therefore, from (3.7) $\sim$ (3.10) we have

$$(3.11)\qquad I_\lambda = \frac{1}{2}\,\bar E[\tilde P_{\lambda\beta_+}\{\gamma_\lambda^{-1}(\Gamma_\lambda)\}] + o(1),\quad \lambda\to\infty\ ,$$

and a similar formula for II_λ.

To complete the proof of the theorem we need the following simple lemma.

<u>*Lemma 4*</u>. *For any $\lambda>0$ and $\Gamma\in\mathcal{B}$ we have*

$$\tilde P_{\lambda\beta_+}\{\gamma_\lambda^{-1}(\Gamma)\} \overset{d}{=} \tilde P_{\beta_+}\{\Gamma\}\ .$$

Proof. For each W with $e^{-W}\in L^1(\mathbb{R})$, $\tilde X(t,W)$ denotes the stationary $\mathcal{L}_W$-diffusion process with initial distribution $\tilde\mu_W$. Fixing a sample path $\beta_+(\cdot)$, we put $\tilde X(t) = \tilde X(t,\beta_+(\cdot))$ and $\tilde X_\lambda(t) = \tilde X(t,\beta_+(\lambda^2\cdot))$. Since the scaling relation (3.5) implies

$$\tilde P_{\lambda\beta_+}\{\gamma_\lambda^{-1}(\Gamma)\} \overset{d}{=} \tilde P_{\beta_+(\lambda^2\cdot)}\{\gamma_\lambda^{-1}(\Gamma)\}\ ,$$

for the proof of the lemma it is enough to show

$$\{\gamma_\lambda\tilde X_\lambda(t),\ t\ge 0\} \overset{d}{=} \{\tilde X(t),\ t\ge 0\}$$

for each fixed sample path $\beta_+(\cdot)$. In what follows the notation Ω

stands for the image measure of $\tilde{\mu}_{\beta_+(\cdot)}$ under the map $\hat{S}: \mathbb{R} \to \mathbb{R}$, where $\hat{S}(x) = \int_0^x e^{\beta_+(y)}dy$. Similarly $\hat{\mu}_\lambda$ stands for the image measure obtained by replacing $\beta_+(\cdot)$ by $\beta_+(\lambda^2\cdot)$. The following fact can be easily verified:

(3.12) If $B(0)$ is a random variable with distribution $\hat{\mu}_\lambda$, then $\lambda^2 B(0)$ is distributed according to $\hat{\mu}$.

As in 1.1, $\tilde{X}_\lambda(t)$ can be constructed from a Brownian motion $B(t)$ with initial distribution $\hat{\mu}_\lambda$. Write $B(t) = B(0) + \tilde{B}(t)$ and put $B_\lambda(t) = B(0) + \lambda^{-2}\tilde{B}(\lambda^4 t)$. Then $B_\lambda(t)$ is again a Brownian motion with initial distribution $\hat{\mu}_\lambda$. Therefore if, in the construction of $\tilde{X}_\lambda(t)$, we use $B_\lambda(t)$ instead of $B(t)$, we still have a diffusion process $\hat{X}_\lambda(t)$ which is equivalent in law to $\tilde{X}_\lambda(t)$. By an easy calculation we see that

$$\hat{X}_\lambda(t) = \lambda^{-2}\hat{S}^{-1}(\lambda^2 B(0) + \tilde{B}(\hat{A}^{-1}(\lambda^4 t))) ,$$

$$= \lambda^{-2}\hat{S}^{-1}(\hat{B}(\hat{A}^{-1}(\lambda^4 t))) ,$$

where $\hat{B}(t) = \lambda^2 B(0) + \tilde{B}(t)$ is a Brownian motion with initial distribution $\hat{\mu}$ (by (3.12)) and $\hat{A}^{-1}(t)$ is the inverse function of

$$\hat{A}(s) = \int_0^s e^{-2\beta_+(\hat{S}^{-1}(\hat{B}(u)))}du .$$

Therefore, $\{\gamma_\lambda \hat{X}_\lambda(t)\}$ and consequently $\{\gamma_\lambda \tilde{X}_\lambda(t)\}$ is equivalent in law to $\{\tilde{X}(t)\}$.

The proof of the theorem is now completed as follows. Take an arbitrary $\Gamma \in \mathcal{B}_t$, $t > 0$ being arbitrarily fixed, and put $\Gamma_\lambda = \Gamma$ in (3.2). Then from (3.2), (3.11) and Lemma 4 we have

$$\lim_{\lambda\to\infty} E_Q[P_W\{\theta_{\exp\lambda}\omega - \lambda^2 b(W_\lambda) \in \Gamma\}] = \bar{E}[\tilde{P}_{\beta_+}\{\Gamma\}] ,$$

and this prove the theorem.

REFERENCES

[1] Th. Brox, A one-dimensional diffusion process in a Wiener medium, Ann. Probab., 14(1986), 1206-1218.

[2] A. O. Golosov, Localization of random walks in one-dimensional random environments, Commun. Math Phys., 92(1984), 491-506.

[3] N. Ikeda and S. Watanabe, Stochastic Differential Equations and Diffusion Processes, North-Holland, 1981.

[4] K. Itô and H. P. McKean, Diffusion Processes and Their Sample Paths, Springer-Verlag, 1965.

[5] K. Itô, Poisson point processes attached to Markov Processes, Proc. 6th Berkeley Symp. Math. Statist. Probab. III, 225-239, Univ. California Press, Berkeley, 1972.

[6] K. Kawazu, Y. Tamura and H. Tanaka, One-dimensional diffusions and random walks in random environments, to appear in Proc. 5th Japan-USSR Symp. Probab. Th.

[7] H. Kesten, The limit distribution of Sinai's random walk in random environment, Physica, 138A(1986), 299-309.

[8] P. Lévy, Processus stochastiques et mouvement brownien, Gauthier-Villars, Paris, 1948.

[9] J. W. Pitman, One-dimensional Brownian motion and the three-dimensional Bessel process, Adv. Appl. Probab., 7(1975), 511-526.

[10] J. W. Pitman and M. Yor, A decomposition of Bessel bridges, Z. Wahrscheinlichkeitstheorie verw. Gebiete, 59(1982), 425-457.

[11] Y. G. Sinai, The limiting behavior of a one-dimensional random walk in a random medium, Theory of Probab. Appl., 27(1982), 256-268.

[12] H. Tanaka, Limit distributions for one-dimensional diffusion processes in self-similar random environments, to appear in Hydrodynamic Behavior and Interacting Particle Systems, the IMA Volumes in Math. and its Appl., Vol. 9, 1987, Springer-Verlag.

[13] D. Williams, Path decomposition and continuity of local time for one-dimensional diffusions, I, Proc. London Math. Soc., (3)28(1974), 738-768.

DIFFUSION PROCESSES AND HEAT KERNELS ON CERTAIN NILPOTENT GROUPS

Hideaki UEMURA and Shinzo WATANABE
Department of Mathematics. Faculty of Science
Kyoto University. Kyoto, 606 (Japan)

Introduction.

Many works have been done on the short time asymptotic behavior of heat kernels by probabilistic methods, cf. e.g. Molchanov [9], Azencott [2] for an approach by pinned diffusion processes, Bismut [4], Kusuoka-Stroock [7], Watanabe [12], Léandre [8], Ben Arous [3], Takanobu [10], Uemura [11] for an approach by the Malliavin calculus. When we consider the short time asymptotic of heat kernel $p(t,x,y)$ off the diagonal, the set $H^{x,y}$ of horizontal curves connecting x and y in the unit time with the minimal action plays an important role (see Section 1 below for the precise definition of these notions.). This fact was first noticed by Molchanov [9] for a non-degenerate heat kernel, i.e. heat kernel associated with the Laplacian on a Riemannian manifold, and was recently studied thoroughly by Kusuoka through his ingenious method of generalized Malliavin calculus. Unfortunately his work is not yet published.

Purpose of the present paper is to study a similar problem for heat kernels associated with semi-elliptic second order differential operators of Hörmander type

$$L = \frac{1}{2}\sum_{\alpha=1}^{r} V_\alpha^{\,2}$$

by a probabilistic method which is based on the Malliavin calculus.

Here we would illustrate our method only in the following concrete cases : $\{ V_\alpha \}_{\alpha=1}^{r}$ generates a free nilpotent Lie algebra of step 2; $r=2$ (the case of Heisenberg group H_3) and $r=4$ (the case of the nilpotent group $N_{4,2}$). In these cases, heat kernels can be explicitely expressed and the short time asymptotics can be computed directly through this expression (cf. Gaveau [5], Azencott [1]). We hope, however, that our approach will provide us with an insight and methods in more general cases of heat kernels associated with semi-elliptic second order differential operators.

We acknowledge that this work has been motivated by the lecture of Professor G.Ben Arous at the French-Japanese Seminar.

§ 1 General notions and notations.

Let V_1 , V_2 , $\cdots$, V_r be C^∞-vector fields on $\mathbb{R}^d$ with C^∞-coefficients whose derivatives of order ≥ 1 are all bounded. Let L be the semi-elliptic operator defined by

$$(1.1) \qquad L = \frac{1}{2} \sum_{\alpha=1}^{r} V_\alpha{}^2 .$$

If the initial value problem of the associated heat equation

$$(1.2) \qquad \begin{cases} \dfrac{\partial u}{\partial t} = Lu \\ u|_{t=0} = f \end{cases}$$

has solution $u(t,x)$ represented in the form

$$u(t,x) = \int_{\mathbb{R}^d} p(t,x,y) f(y) dy \quad ,$$

dy being the d-dimensional Lebesgue measure, the kernel $p(t,x,y)$ is called the *fundamental solution* or the *heat kernel* associated with

(1.2).

Let (W,P) be the r-dimensional Wiener space with the time interval $[0,1]$; So, W is a real Banach space formed of continuous paths $w : [0,1] \ni t \longrightarrow w(t) \in \mathbb{R}^r$ such that $w(0)=0$ endowed with $\|w\| = \max_{0 \le t \le 1} |w(t)|$ and P is the standard r-dimensional Wiener measure on W. Let $H \subset W$ be a Hilbert subspace (called the *Cameron-Martin space*) formed of $h \in W$ such that h is absolutely continuous with square-integrable derivative endowed with the Hilbertian norm $\|h\|_H = \left(\int_0^1 |\frac{dh}{dt}|^2 dt \right)^{1/2}$. Given the above vector fields $\{V_\alpha\}$ and semi-elliptic operator L , there is associated *the L-diffusion process* which can be realized on the Wiener space by the solution $X(t) = X(t,x,w)$ of the stochastic differential equation (S.D.E.)

$$(1.3) \qquad \begin{cases} dX(t) = \sum_{\alpha=1}^{r} V_\alpha(X(t)) \circ dw^\alpha(t) \\ X(0) = x \in \mathbb{R}^d \end{cases}$$

where $w(t) = (w^\alpha(t)) \in W$ and $\circ$ denotes the stochastic differential in the Stratonovich sense. More generally, it is convenient to introduce the following S.D.E. with parameter $\varepsilon \in (0,1]$:

$$(1.4) \qquad \begin{cases} dX(t) = \varepsilon \sum_{\alpha=1}^{r} V_\alpha(X(t)) \circ dw^\alpha(t) \\ X(0) = x \end{cases}$$

The solution of (1.4) is denoted by $X^\varepsilon(t,x,w)$. It is immediately seen that $\{X^\varepsilon(t,x,w)\} \overset{\mathcal{L}}{\sim} \{X(\varepsilon^2 t,x,w)\}$.

A continuous curve $c(t)$ in $\mathbb{R}^d$ is called *horizontal with respect to* $\{V_\alpha\}$ if it is absolutely continuous in t and satisfies

$$(1.5) \qquad \frac{dc(t)}{dt} = \sum_{\alpha=1}^{r} V_\alpha(c(t)) \frac{dh^\alpha}{dt}(t) , \qquad 0 \le t \le 1 ,$$

for some $h = (h^{\alpha}(t)) \in H$. Given $x \in \mathbb{R}^d$ and $h \in H$, there exist a unique horizontal curve $c(t) := c^{x,h}(t)$ satisfying (1.5) and $c(0) = x$. Given x and y in $\mathbb{R}^d$, let

$$(1.6) \qquad K^{x,y} = \{ h \in H ; c^{x,h}(1) = y \} .$$

In the following, unless otherwise stated, we assume that the following condition (H.1) *is satisfied at every* $x \in \mathbb{R}^d$:

$$(\text{H.1}) \qquad \dim \mathcal{L}ie \{ V_1 , V_2 , \cdots , V_r \}_x = d .$$

Here $\mathcal{L}ie \{ V_1 , \cdots , V_r \}_x$ is the subspace of $\mathbb{R}^d$ ($= T_x(\mathbb{R}^d)$) obtained by restricting at x of the Lie subalgebra $\mathcal{L}ie \{ V_1 , \cdots \cdots , V_r \}$ of vector fields generated by $\{ V_\alpha \}$. Under this assumption, it is well-known that

$$(1.7) \qquad K^{x,y} \neq \phi \qquad \text{for every } x , y \in \mathbb{R}^d .$$

Also it is well-known that the Malliavin covariance of $X^{\varepsilon}(1,x,w) \in \mathbb{D}^{\infty}(\mathbb{R}^d)$ is non-degenerate in the sense of Malliavin for each $\varepsilon \in (0,1]$ (Cf. [12] for notions and notations concerning the Malliavin calculus), and hence, if δ_y is the Dirac δ-function at $y \in \mathbb{R}^d$, $\delta_y(X^{\varepsilon}(1,x,w)) \in \tilde{\mathbb{D}}^{-\infty}(\mathbb{R}^d)$ is well-defined as a generalized Wiener functional. Then we can define

$$(1.8) \qquad p(\varepsilon^2,x,y) = E[\delta_y(X^{\varepsilon}(1,x,w))]$$

by the generalized Wiener functional expectation and it turns out that $p(t,x,y)$ is the heat kernel associated with (1.2). Thus we see, in particular, that the above assumption is sufficient to guarantee the existence of the heat kernel for (1.2). Furthermore, we can study many properties of $p(t,x,y)$, regularities and asymptotics, in particular, through the expression (1.8), some examples of which we will discuss below.

Given $x \in \mathbb{R}^d$ and $h \in H$, consider the equation for $c(t) =$

$(c^i(t))$ and $Y(t) = (Y^i_j(t))$, $i,j=1,2,\cdots,d$, written in the matrix notation :

$$(1.9)\quad \begin{cases} \dfrac{dc(t)}{dt} = \sum_{\alpha=1}^{r} V_\alpha(c(t))\dfrac{dh^\alpha}{dt}(t) \\ c(0) = x \\ \dfrac{dY(t)}{dt} = \sum_{\alpha=1}^{r} \partial V_\alpha(c(t))Y(t)\dfrac{dh^\alpha}{dt}(t) \\ Y(0) = I \quad (\text{ = identity matrix }) \end{cases}$$

where $\partial V_\alpha(x) = \left(\dfrac{\partial V^i_\alpha(x)}{\partial x^j} \right)$, $i,j=1,2,\cdots,d$. Then clearly $c(t) = c^{x,h}(t)$ and, we denote the solution $Y(t)$ by $Y^{x,h}(t)$. Let $d\times d$-matrix $\Xi^{x,h}$ be defined by

$$(1.10)\quad \Xi^{x,h} = \sum_{\alpha=1}^{r} \int_0^1 Y^{x,h}(1)Y^{x,h}(t)^{-1}V_\alpha(c^{x,h}(t)) \otimes Y^{x,h}(1)Y^{x,h}(t)^{-1}V_\alpha(c^{x,h}(t))dt \ .$$

$\Xi^{x,h}$ is called the *deterministic Malliavin covariance* for x and h.

For x , $y \in \mathbb{R}^d$, $x\neq y$, let

$$(1.11)\quad d(x,y) = \min \{ \|h\|_H \ ; \ h \in K^{x,y} \}$$

and call it the *control metric* of x and y . Let

$$K^{x,y}_{\min} = \{ h \in K^{x,y} \ ; \ \|h\| = d(x,y) \} \ .$$

Then it is known that $K^{x,y}_{\min} \neq \phi$. We define $M^{x,y}$ by

$$(1.12)\quad M^{x,y} = \{ c^{x,h} \ ; \ h \in K^{x,y}_{\min} \}$$

and call it the set of all horizontal curves connecting x and y in the unit time with the minimal action.

We define the *Hamiltonian* $H(x,p)$ associated with the vector fields $\{ V_\alpha \}$ by

$$(1.13)\quad H(x,p) = \frac{1}{2} \sum_{\alpha=1}^{r} \langle p,V_\alpha(x)\rangle^2 \ , \ (x,p) \in \mathbb{R}^d\times\mathbb{R}^d \ (= T^*(\mathbb{R}^d) \ ) \ ,$$

where $\langle\ ,\ \rangle$ is the $\mathbb{R}^d$-inner product (more intrinsically, the pairing of $T_x^*(\mathbb{R}^d)$ and $T_x(\mathbb{R}^d)$). Consider the following Hamilton-Jacobi equation

$$(1.14)\qquad \begin{cases} \dfrac{dx}{dt}(t) = \dfrac{\partial H}{\partial p}(x(t),p(t)) \\ \dfrac{dp}{dt}(t) = -\dfrac{\partial H}{\partial x}(x(t),p(t)) \end{cases} .$$

A solution $(x(t),p(t))$ of (1.14) is called a *bicharacteristic* . Some of known facts are in order ; (cf. Bismut[4])

(1) The projection $x(t)$ of a bicharacteristic $(x(t),p(t))$ is a horizontal curve : Indeed it is associated with $h = (h^\alpha(t))$ given by $\dfrac{dh^\alpha}{dt}(t) = \langle p(t),V_\alpha(x(t))\rangle$.

(2) If $h \in K_{\min}^{x,y}$ and the deterministic Malliavin covariance $\Xi^{x,h}$ is non-degenerate, i.e. $\det \Xi^{x,h} > 0$, then there exists a unique $p \in \mathbb{R}^d$ $(= T_x^*(\mathbb{R}^d))$ such that $\dfrac{dh^\alpha}{dt}(t) = \langle p, Y^{x,h}(1)Y^{x,h}(t)^{-1}V_\alpha(c^{x,h}(t))\rangle$, $0\le t\le 1$, and $c^{x,h}(t)$ is the projection of the bicharacteristic starting at (x,p).

(3) The deterministic Malliavin covariance $\Xi^{x,h}$ is non-degenerate for every $h \in H$, $h \neq 0$, if the following condition (H.2) is satisfied at x :

(H.2) $V_1(x), \cdots , V_r(x)$ are linearly independent and, for every $\lambda \in \mathbb{R}^r$, $\lambda\neq 0$,

$$\dim \textit{linear span} \{ V_1(x), \cdots , V_r(x), [V_1,Y](x), \cdots , [V_r,Y](x) \} = d$$

where $Y = \sum_{\alpha=1}^{r} \lambda_\alpha V_\alpha(x)$, $\lambda=(\lambda_1,\cdots,\lambda_r)$.

In this paper, we mainly consider the following two cases of vector fields :

<u>The case 1</u> . $d=3$, $r=2$ and V_1 and V_2 are given by

$$(1.15)\qquad \begin{cases} V_1 = \dfrac{\partial}{\partial x_1} + 2x_2 \dfrac{\partial}{\partial x_3} \\ V_2 = \dfrac{\partial}{\partial x_2} - 2x_1 \dfrac{\partial}{\partial x_3} \end{cases} \qquad x = (x_1, x_2, x_3) \in \mathbb{R}^3 .$$

The case 2 . $d=10$, $r=4$ and, introducing the coordinate of $x \in \mathbb{R}^{10}$ by $x = (x_1, x_2, x_3, x_4, x_{12}, x_{13}, x_{14}, x_{23}, x_{24}, x_{34})$,

$$(1.16)\qquad V_i = \frac{\partial}{\partial x_i} + \frac{1}{2}\Big(\sum_{\substack{1\le k\le 4\\ k<i}} x_k \frac{\partial}{\partial x_{ki}} - \sum_{\substack{1\le k\le 4\\ k>i}} x_k \frac{\partial}{\partial x_{ik}} \Big) \qquad i=1,2,3,4.$$

These cases are canonical realizations of free nilpotent Lie algebras of step 2 with 2 and 4 generators, respectively. In the first case, the corresponding Lie group H_3 is realized by $\mathbb{R}^3$ with group operation

$$(x_1, x_2, x_3)\cdot(y_1, y_2, y_3) = (x_1+y_1, x_2+y_2, x_3+y_3+2(x_1y_2-x_2y_1)) .$$

This group $\mathbb{R}^3$ is called the *Heisenberg group* . In the second case, the corresponding Lie group $N_{4,2}$ is realized by $\mathbb{R}^{10}$ with group operation

$$(x_i, x_{ij})\cdot(y_i, y_{ij}) = (x_i+y_i, x_{ij}+y_{ij}+\tfrac{1}{2}(x_iy_j-x_jy_i)).$$

§ 2 The case of Heisenberg group.

We consider, on $\mathbb{R}^3$, the vector fields V_1 and V_2 given by (1.15). Then

$$(2.1)\qquad L = \frac{1}{2}\Big(\frac{\partial^2}{\partial x_1^2} + \frac{\partial^2}{\partial x_2^2} + 4x_2\frac{\partial^2}{\partial x_1\partial x_3} - 4x_1\frac{\partial^2}{\partial x_2\partial x_3} + 4(x_1^2+x_2^2)\frac{\partial^2}{\partial x_3^2}\Big)$$

and the L-diffusion is given on the 2-dimensional Wiener space (W,P) by

$$(2.2)\qquad \begin{cases} X^1(t,x,w) = x_1 + w^1(t) \\ X^2(t,x,w) = x_2 + w^2(t) \\ X^3(t,x,w) = x_3 + 2[\, x_2w^1(t) - x_1w^2(t) \,] + 2S(t,w) \end{cases}$$

where $S(t,w)$ is the stochastic area integral of P.Lévy for two dimensional Brownian motion $w(t) = (w^1(t),w^2(t))$:

$$S(t,w) = \int_0^t w^2(s)dw^1(s) - w^1(s)dw^2(s) \quad .$$

Since $[V_1,V_2] = V_1V_2 - V_2V_1 = -4\frac{\partial}{\partial x_3}$, not only (H.1) but also (H.2) are satisfied at every $x \in \mathbb{R}^3$. In particular, the fundamental solution $p(t,x,y)$ exists. We are interested in the short time asymptotic behavior of $p(t,x,y)$ as $t\searrow 0$ when $x\neq y$. Since $p(t,x,y)$ has an invariance under the group action, we may assume $x = 0 \in \mathbb{R}^3$. We know that the set $M^{0,x}$, $x\neq 0$, of horizontal curves connecting 0 and x in the unit time with the minimal action can be obtained by projecting bicharacteristics starting at $(0,p)$, $p \in \mathbb{R}^3$. Now the Hamiltonian (1.13) is given by

$$H(x,p) = \frac{1}{2}(p_1{}^2 + p_2{}^2 + 4x_2p_1p_3 - 4x_1p_2p_3 + 4(x_1{}^2+x_2{}^2)p_3{}^2)$$

and hence, the Hamilton-Jacobi equation is given by (denoting $\dot{} = \frac{d}{dt}$)

$$\begin{aligned} \dot{x}_1 &= p_1 + 2x_2p_3 \\ \dot{x}_2 &= p_2 - 2x_1p_3 \\ \dot{x}_3 &= 2x_2p_1 - 2x_1p_2 + 4(x_1{}^2+x_2{}^2)p_3 \\ \dot{p}_1 &= 2p_2p_3 - 4x_1p_3{}^2 \\ \dot{p}_2 &= -2p_1p_3 - 4x_2p_3{}^2 \\ \dot{p}_3 &= 0 \quad . \end{aligned}$$

Therefore, the bicharacteristic starting at $(\, 0,0,0,2\beta\sigma,2\alpha\sigma,\frac{\sigma}{2}\,)$, $\sigma\neq 0$, is given by

$$(2.3)\quad \begin{cases} x_1(s) = \alpha(1-\cos 2\sigma s) + \beta\sin 2\sigma s \\ x_2(s) = \alpha\sin 2\sigma s - \beta(1-\cos 2\sigma s) \\ x_3(s) = 2(\alpha^2+\beta^2)(2\sigma s - \sin 2\sigma s) \\ p_1(s) = \alpha\sigma\sin 2\sigma s + \beta\sigma(1+\cos 2\sigma s) \\ p_2(s) = \alpha\sigma(1+\cos 2\sigma s) - \beta\sigma\sin 2\sigma s \\ p_3(s) = \sigma/2 \end{cases}$$

and the bicharacteristic starting at $(0,0,0,\alpha,\beta,0)$ is given by

$$(2.4)\quad \begin{cases} x_1(s) = \alpha s , & p_1(s) = \alpha \\ x_2(s) = \beta s , & p_2(s) = \beta \\ x_3(s) = 0 , & p_3(s) = 0 . \end{cases}$$

Clearly, the curve $x(s) = (x_1(s),x_2(s),x_3(s))$ given by (2.3) or (2.4) is the horizontal curve $c^{0,h}$ with $h(s) = (x_1(s),x_2(s))$. Thus in order to obtain $K^{0,x}_{\min}$, we have only to determine the parameters α,β,σ so that $x(1) = x$ and $\|h\|_H$ is minimal. If $x_1{}^2+x_2{}^2 > 0$, then it is easy to see (cf. [5] or [1]) that $K^{0,x}_{\min}$ consists of just one element $\bar{h}$ while if $x_1{}^2+x_2{}^2 = 0$ and $x_3\neq 0$, say $x_3 > 0$ for simplicity, then $K^{0,x}_{\min} = \{ h_\theta ; \theta \in [0,2\pi) \}$ where, by setting $r = (x_3/4\pi)^{1/2}$,

$$(2.5)\quad \begin{cases} h^1_\theta(t) = r\sin\theta(1-\cos 2\pi t) + r\cos\theta\sin 2\pi t \\ h^2_\theta(t) = r\sin\theta\sin 2\pi t - r\cos\theta(1-\cos 2\pi t) . \end{cases}$$

Now we compute the asymptotics of $p(t,0,x)$ as $t\searrow 0$. First we consider the case $x_1{}^2+x_2{}^2 > 0$. Then $x(s) = (x_1(s),x_2(s),x_3(s))$ in (2.3) or (2.4) with $x(1) = x$ is uniquely determined and the above $\bar{h}$ is given by $(x_1(s),x_2(s))$. In the following, we study mainly the case $x_3\neq 0$, i.e. the case of (2.3); the case $x_3=0$ being similarly discussed with obvious modification. Then α,β,σ corresponding to $\bar{h}$ satisfy

$$(2.6)\quad x_1{}^2+x_2{}^2 = 4(\alpha^2+\beta^2)\sin^2\sigma ,$$

$$(2.7)\quad \frac{2\sigma - \sin 2\sigma}{2\sin^2\sigma} = \frac{x_3}{x_1{}^2+x_2{}^2} ,$$

(2.8) $$\|\bar{h}\|_H^2 = 4(\alpha^2+\beta^2)\sigma^2 = (x_1{}^2+x_2{}^2)\left(\frac{\sigma}{\sin\sigma}\right)^2 .$$

Now

$$p(\varepsilon^2,0,x) = E[\delta_x(X^\varepsilon(1,0,w))]$$
$$= E[\delta_x(\varepsilon w^1(1),\varepsilon w^2(1),2\varepsilon^2 S(1,w))]$$

and, by the Cameron-Martin theorem applied to the translation $w \longrightarrow w + \bar{h}/\varepsilon$, the above is equal to

$$E[\exp(-\|\bar{h}\|_H^2/2\varepsilon^2 - (\bar{h},w)_H/\varepsilon)$$
$$\delta_x(x_1+\varepsilon w^1(1),x_2+\varepsilon w^2(1),x_3+4\varepsilon[\int_0^1 \bar{h}^2(s)dw^1(s)-\bar{h}^1(s)dw^2(s)]$$
$$+2\varepsilon[w^2(1)x_1-w^1(1)x_2]+2\varepsilon^2 S(1,w))]$$
$$= \varepsilon^{-3} E[\exp(-\|\bar{h}\|_H^2/2\varepsilon^2 - (\bar{h},w)_H/\varepsilon)$$
$$\delta_0(w^1(1),w^2(1),4[\int_0^1 \bar{h}^2(s)dw^1(s)-\bar{h}^1(s)dw^2(s)]+2\varepsilon S(1,w))]$$

where we usually denote by $(h,w)_H$, $h \in H$, the Wiener integral $\sum_{i=1}^2 \int_0^1 h^i(s)dw^i(s)$. Under the condition $w^1(1)=w^2(1)=0$,

$$(\bar{h},w)_H = 2\sigma[\int_0^1 \bar{h}^2(s)dw^1(s) - \int_0^1 \bar{h}^1(s)dw^2(s)] .$$

Thus under the condition $f_\varepsilon(w) = 0$ where $f_\varepsilon(w) = (w^1(1),w^2(1),$ $4[\int_0^1 \bar{h}^2(s)dw^1(s)-\bar{h}^1(s)dw^2(s)] + 2\varepsilon S(1,w))$, $(\bar{h},w)_H = -\varepsilon\sigma S(1,w)$.

Hence we obtain

$$p(\varepsilon^2,0,x) = \varepsilon^{-3}\exp(-\|\bar{h}\|_H^2/2\varepsilon^2)E[e^{\sigma S(1,w)}\delta_0(f_\varepsilon)\varphi(w(1))]$$

where φ is a C^∞-function on $\mathbb{R}^2$ with a compact support such that $\varphi=1$ on $|x|\le 1$. Since $|\sigma|<\pi$, σ being the unique solution in $(-\pi,\pi)$ of (2.7), $e^{\sigma S(1,w)}\varphi(w(1)) \in \tilde{\mathbb{D}}^{-\infty}$. On the other hand, it is easy to see that $f_\varepsilon \in \mathbb{D}^\infty(\mathbb{R}^3)$ is non-degenerate in the sense of Malliavin uniformly in $\varepsilon \in [0,1)$ and $f_0 = (w^1(1),w^2(1),$ $4[\int_0^1 \bar{h}^2(s)dw^1(s)-\bar{h}^1(s)dw^2(s)])$ is a non-degenerate Gaussian random variable. If C is the covariance (= the Malliavin covariance) of f_0 , it is easy to see that

$$\det C = 16(\alpha^2+\beta^2)\left(\frac{\sigma^2-\sin^2\sigma}{\sigma^2}\right) = 4(x_1{}^2+x_2{}^2)\left(\frac{\sigma^2-\sin^2\sigma}{\sigma^2\sin^2\sigma}\right) .$$

Then we can apply a general result of [12] to conclude that

$$p(\varepsilon^2,0,x) \sim \varepsilon^{-3}\exp(-\|\bar{h}\|_H^2/2\varepsilon^2)E[e^{\sigma S(1,w)}\delta_0(f_0)] .$$

The expectation

$$E[e^{\sigma S(1,w)}\delta_0(f_0)] = E[e^{\sigma S(1,w)}|f_0=0][(2\pi)^3\det C]^{-1/2}$$

can be explicitely computed as in [4], pp.204 ~ 207, to obtain, by writing $\varepsilon^2=t$,

$$(2.9) \qquad p(t,0,x) \sim (2\pi t)^{-3/2}\exp\{-\frac{(x_1{}^2+x_2{}^2)}{2t}\left(\frac{\sigma}{\sin\sigma}\right)^2\}\times \frac{\sigma}{2}\left(\frac{\sin\sigma}{\sin\sigma-\sigma\cos\sigma}\right)^{1/2}(x_1{}^2+x_2{}^2)^{-1/2} .$$

Next, we consider the case $x_1{}^2+x_2{}^2=0$. A striking feature in this case is that $K_{\min}^{0,x}$ does not consists of a single element but constitutes a one parameter family of elements h_θ , $\theta \in [0,2\pi)$. As a consequence, the power of t in (2.9) changes from -3/2 to -2. Namely we have the following result:

$$(2.10) \qquad p(t,0,(0,0,x_3)) \sim \frac{1}{4t^2}\exp\{-\pi x_3/2t\} , \qquad \text{as} \quad t\searrow 0 .$$

We will prove (2.10) by the following probabilistic method.

<u>Remark</u>. A formal computation given in [12], pp.36 ~ 37, was wrong. It was based on the following intuitive argument : using the notations given below,

$$\exp\{\pi S(1,w)\}\chi(g_{\varepsilon,\theta}(w)) = \exp\{\pi S(1,w)\}\chi(g_{0,\theta}(w)) + O(\varepsilon) \quad \text{in } \tilde{\mathbb{D}}^{-\infty} \text{ as } \varepsilon\searrow 0$$

but *this asymptotic is false*. The formula

$$p(\varepsilon^2,0,(0,0,x_3)) \sim \frac{1}{8\varepsilon^4}\exp(-\pi x_3/2\varepsilon^2)$$

is therefore wrong and should be corrected as (2.10). Note that there is the same mistake of constant in [5] and [1].

Let χ be a C^∞-function on $\mathbb{R}^4$ with a compact support such that $\chi \equiv 1$ on $|x|\leq 1$. For $\varepsilon \in [0,1]$ and $\theta \in [0,2\pi)$, set

$$(2.11)\qquad g_{\varepsilon,\theta}(w) = (w^1(1),w^2(1),(h_\theta,w)_H+\varepsilon\pi S(1,w),(h_{\theta-(\pi/2)},w)_H) \in \mathbb{D}^\infty(\mathbb{R}^4)\ .$$

In particular,

$$(2.12)\qquad g_{0,\theta}(w) = (w^1(1),w^2(1),(h_\theta,w)_H,(h_{\theta-(\pi/2)},w)_H)\ .$$

Lemma 2.1

$\exp(\pi S(1,w))\cdot\chi(g_{0,\theta}(w)) \in \tilde{\mathbb{D}}^\infty$ *for every* $\theta \in [0,2\pi)$ *and*

$$(2.13)\qquad E[\exp(\pi S(1,w))\delta_{(0,0,0,0)}(g_{0,\theta}(w))] = \frac{1}{16\pi^4 r^2}\ .$$

Note that this generalized expectation is well-defined as the coupling of $\exp(\pi S(1,w))\cdot\chi(g_{0,\theta}(w)) \in \tilde{\mathbb{D}}^\infty$ *and* $\delta_{(0,0,0,0)}(g_{0,\theta}(w)) \in \tilde{\mathbb{D}}^{-\infty}$ *and it coincides with*

$$E[\exp(\pi S(1,w))\,|\,g_{0,\theta}(w)=0]\cdot\frac{1}{(2\pi)^2\sqrt{\det C}}$$

where C *is the covariance (= the Malliavin covariance) of* $g_{0,\theta}(w)$ *and* $\sqrt{\det C} = (2\pi r)^2$.

Proof. Set, for i = 1 and 2 ,

$$\xi_k^{(i)} = \sqrt{2}\int_0^1 \sin 2\pi k t\,dw^i(t)\ ,\quad \eta_k^{(i)} = \sqrt{2}\int_0^1 \cos 2\pi k t\,dw^i(t)\ ,\quad k = 1,2,\cdots\ .$$

Then

$$\pi S(1,w) = \sum_{k=1}^{\infty} \frac{1}{k}(\eta_k^{(2)}\xi_k^{(1)} - \eta_k^{(1)}\xi_k^{(2)}) + \sqrt{2}\left(w^1(1)\sum_{k=1}^{\infty}\frac{\xi_k^{(2)}}{k} - w^2(1)\sum_{k=1}^{\infty}\frac{\xi_k^{(1)}}{k}\right)$$

and

$$\begin{aligned}(h_\theta, w)_H &= \frac{2\pi r}{\sqrt{2}}(\sin\theta\cdot\xi_1^{(1)} + \cos\theta\cdot\eta_1^{(1)} + \sin\theta\cdot\eta_1^{(2)} - \cos\theta\cdot\xi_1^{(2)}) \\ &:= \frac{2\pi r}{\sqrt{2}} F_1(\theta, w) ,\end{aligned}$$

$$\begin{aligned}(h_{\theta-(\pi/2)}, w)_H \\ &= -\frac{2\pi r}{\sqrt{2}}(\cos\theta\cdot\xi_1^{(1)} - \sin\theta\cdot\eta_1^{(1)} + \cos\theta\cdot\eta_1^{(2)} + \sin\theta\cdot\xi_1^{(2)}) \\ &:= -\frac{2\pi r}{\sqrt{2}} F_2(\theta, w) .\end{aligned}$$

Since

$$\begin{pmatrix} F_1 \\ F_2 \end{pmatrix} = \begin{pmatrix} \cos\theta & \sin\theta \\ -\sin\theta & \cos\theta \end{pmatrix}\begin{pmatrix} \eta_1^{(1)} - \xi_1^{(2)} \\ \xi_1^{(1)} + \eta_1^{(2)} \end{pmatrix}$$

the condition $\chi(g_{0,\theta}(w)) > 0$ implies that

$$|w^1(1)| + |w^2(1)| + |\eta_1^{(1)} - \xi_1^{(2)}| + |\xi_1^{(1)} + \eta_1^{(2)}| < \delta$$

for some $\delta > 0$. Now

$$\begin{aligned}\pi S(1,w) &= (\eta_1^{(2)}\xi_1^{(1)} - \eta_1^{(1)}\xi_1^{(2)}) + \sum_{k=2}^{\infty}\frac{1}{k}(\eta_k^{(2)}\xi_k^{(1)} - \eta_k^{(1)}\xi_k^{(2)}) \\ &\quad + \sqrt{2}\left(w^1(1)\sum_{k=1}^{\infty}\frac{\xi_k^{(2)}}{k} - w^2(1)\sum_{k=1}^{\infty}\frac{\xi_k^{(1)}}{k} \right) \\ &:= I_1 + I_2 + I_3\end{aligned}$$

and

$$E[e^{\alpha I_2}] = \prod_{k=2}^{\infty}(1 - \frac{\alpha^2}{k^2})^{-1} < \infty \quad \text{if} \quad |\alpha| < 2 .$$

It is obvious that

$$E[e^{\alpha I_3} ; |w^1(1)| + |w^2(1)| < \delta] < \infty \quad \text{for all} \quad 1 < \alpha < \infty .$$

Also, since

$$I_1 = -(\xi_1^{(1)})^2 - (\xi_1^{(2)})^2 + \xi_1^{(1)}(\xi_1^{(1)} + \eta_1^{(2)}) - \xi_1^{(2)}(\eta_1^{(1)} - \xi_1^{(2)})$$

it is obvious that

$$E[e^{\alpha I_1} ; |\eta_1^{(1)} - \xi_1^{(2)}| + |\xi_1^{(1)} + \eta_1^{(2)}| < \delta] < \infty \text{ for all } 1 < \alpha < \infty.$$

Thus we have obtained that

$$E[\exp(p\pi S(1,w))\cdot\chi(g_{0,\theta}(w))] < \infty \quad \text{for all} \quad 1< p <2 .$$

This clearly implies the first assertion of the lemma. Furthermore,

$$\begin{aligned}
(2.14)\quad & E[\exp\{\pi S(1,w)\}\delta_{(0,0,0,0)}(g_{0,\theta}(w))] \\
& (= E[\exp\{\pi S(1,w)\}\cdot\chi(g_{0,\theta}(w))\cdot\delta_{(0,0,0,0)}(g_{0,\theta}(w))] \ ) \\
& = \frac{1}{(2\pi r)^2} E[\exp\{\pi S(1,w)\} \\
& \qquad \delta_{(0,0,0,0)}(w^1(1),w^2(1),(\eta_1^{(1)}-\xi_1^{(2)})/\sqrt{2},(\xi_1^{(1)}+\eta_1^{(2)})/\sqrt{2})] \\
& = \frac{1}{(2\pi r)^2}\cdot\frac{1}{(2\pi)^2} E[\exp\{\pi S(1,w)\}| \\
& \qquad w^1(1)=0,w^2(1)=0,(\eta_1^{(1)}-\xi_1^{(2)})/\sqrt{2}=0,(\xi_1^{(1)}+\eta_1^{(2)})/\sqrt{2}=0 \ ] \\
& = \frac{1}{16\pi^4r^2} E[\exp\{I_1+I_2\}|(\eta_1^{(1)}-\xi_1^{(2)})/\sqrt{2}=0,(\xi_1^{(1)}+\eta_1^{(2)})/\sqrt{2}=0 \ ] \\
& = \frac{1}{16\pi^4r^2} E[e^{I_1}|(\eta_1^{(1)}-\xi_1^{(2)})/\sqrt{2}=0,(\xi_1^{(1)}+\eta_1^{(2)})/\sqrt{2}=0]\cdot E[e^{I_2}] \\
& = \frac{1}{16\pi^4r^2} E[\exp\{-\tfrac{1}{2}[\{(\eta_1^{(1)}+\xi_1^{(2)})/\sqrt{2}\}^2+\{(\xi_1^{(1)}-\eta_1^{(2)})/\sqrt{2}\}^2]\}] \\
& \qquad\qquad \times E[\exp I_2] \\
& = \frac{1}{16\pi^4r^2}\cdot\frac{1}{2\pi}\left(\int_{-\infty}^{\infty} e^{-x^2}dx\right)^2\cdot\prod_{k=2}^{\infty}\left(1-\frac{1}{k^2}\right)^{-1} = \frac{1}{16\pi^4r^2} . \qquad //
\end{aligned}$$

Let $\delta_0(x)$ be the Dirac δ-function on $\mathbb{R}^1$. Then for each $\theta \in [0,2\pi)$, a distribution $\delta_0(x_1\cos\theta - x_2\sin\theta)$ on $\mathbb{R}^2$ is defined in the usual way (cf.[6],Chap.III). Let $\rho > 0$ and $\bar{\theta} \in [0,2\pi)$ be given. Define $\bar{x} = (\rho\cdot\cos\bar{\theta},\rho\cdot\sin\bar{\theta})$.

Lemma 2.2 *Let $\eta > 0$ satisfy $0 < \eta < \rho$ and $c(\eta) < \pi/4$ where $c(\eta) = \max\{|\arg(x)-\bar{\theta}|;|x-\bar{x}|\le\eta\}$. Then*

$$(2.15)\quad \int_{|\theta-\bar{\theta}|<c(\eta)} \delta_0(x_1\sin\theta-x_2\cos\theta)\cdot(x_1\cos\theta+x_2\sin\theta)d\theta = 1$$

everywhere on $\{x;|x-\bar{x}|<\eta\}$.

Proof is easy and omitted. Note that $c(\eta) \searrow 0$ as $\eta \searrow 0$ and $(x_1\cos\theta+x_2\sin\theta) = |x|\cos(\arg(x)-\theta) > 0$ if $|x-\bar{x}| < \eta$.

To obtain the asymptotic of $p(\varepsilon^2,0,(0,0,x_3))$ as $\varepsilon\searrow 0$ by a probabilistic method, we have to appeal to a large deviation argument as was given in [12] in a similar situation. Since $K := K_{\min}^{0,(0,0,x_3)} = \{h_\theta;\theta \in [0,2\pi)\}$ is compact in W, we can find, for any $\gamma > 0$, a finite $h_{\theta_1}, \cdots, h_{\theta_n} \in K$ such that

$$K \subset \bigcup_{i=1}^{n} V_i$$

where $V_i = \{w \in W;\ \|w-h_{\theta_i}\|_2^2 := |w(1)-h_{\theta_i}(1)|^2 + \int_0^1 |w(s)-h_{\theta_i}(s)|^2 ds < \gamma\}$.

For $w \in W$, we extend the definition of $(w,h_\theta)_H$ by setting

$$(w,h_\theta)_H = \sum_{i=1}^{2} [\ h_\theta^i(1)w^i(1) - \int_0^1 \dot{h}_\theta^{\,i}(s)w^i(s)ds\ ]\ .$$ Clearly

$|(w,h_\theta)_H - (w',h_\theta)_H| \le \kappa\|w-w'\|_2$ for some constant κ independent of θ. In the following, we assume that $\gamma > 0$ is given and then determine θ_i, $i=1,2,\cdots,n=n(\gamma)$ as above but this γ will be chosen sufficiently small later. Let

$$U_i = \{\ w\ ;\ \|w-h_{\theta_i}\|_2^2 < 2\gamma\ \} \supset V_i\ ,\quad i=1,2,\cdots,n\ .$$

Let $\psi(\xi)$ be a C^∞-function on $\mathbb{R}$ such that $0 \le \psi \le 1$, $\psi(\xi)=1$ if $|\xi|<\gamma$ and $\psi(\xi)=0$ if $|\xi|\ge 2\gamma$. Let $\Psi_i(w) = \psi(\|h_{\theta_i}-w\|_2^2)$. Then

$$I_{U_i}(w) \ge \Psi_i(w) \ge I_{V_i}(w)\ .$$

Set $\Phi(w) = 1 - \prod_{i=1}^{n}(1-\Psi_i(w))$. Then

$$1 - \Phi(w) \le I_{\bigcap_{i=1}^{n} V_i^c}$$

and $\bigcap_{i=1}^{n} V_i^c$ is an closed set which is disjoint from K. Clearly $1-\Phi(\varepsilon w) \in \mathbb{D}^\infty$ and also it is continuous on W. Now

$$p(\varepsilon^2,0,(0,0,x_3)) = E[\delta_{(0,0,x_3)}(\varepsilon w^1(1),\varepsilon w^2(1),2\varepsilon^2 S(1,w))]$$
$$= E[\delta_{(0,0,x_3)}(\varepsilon w^1(1),\varepsilon w^2(1),2\varepsilon^2 S(1,w))(1-\Phi(\varepsilon w))]$$
$$+ E[\delta_{(0,0,x_3)}(\varepsilon w^1(1),\varepsilon w^2(1),2\varepsilon^2 S(1,w))\Phi(\varepsilon w)]$$

$= J_1 + J_2$.

We can deduce by an integration by parts on the Wiener space and a standard large deviation argument that

(2.16) $\quad J_1 = O(\exp\{-\pi x_3(1+c)/2\varepsilon^2\})$ *for some* $c > 0$.

Cf. [12], Lemma 3.3 for this type of argument.

We have

$$\Phi = 1 - \prod_{i=1}^{n}(1-\Psi_i) = \sum_{i=1}^{n}\Phi_i$$

where $\Phi_1 = \Psi_1$, $\Phi_2 = \Psi_2(1-\Psi_1)$, $\Phi_3 = \Psi_3(1-\Psi_1)(1-\Psi_2)$, $\cdots$. Then clearly $\Phi_i \cdot I_{U_i} = \Phi_i$, $i=1,\cdots,n$. For $w \in W$, we define $x(w) \in \mathbb{R}^2$ by $x(w) = ((w,h_0/2\pi r)_H, (w,h_{\pi/2}/2\pi r)_H)$. We choose γ sufficiently small that $w \in U_i$ implies $|x(w)-x(h_{\theta_i})| < \eta$ and η satisfies the condition of Lemma 2.2 for $\bar{x} = x(h_{\theta_i})$ (hence $\rho = |x(h_{\theta_i})| = 2\pi r$ and $\bar{\theta} = \theta_i$). Then we have

(2.17)
$$\int_{|\theta-\theta_i|<c(\eta)} \delta_0((h_{\theta-(\pi/2)},w)_H)(h_\theta,w)_H d\theta\cdot\Phi_i(w) = \Phi_i(w) ,$$
$$w \in W , \quad i=1,2,\cdots,n .$$

Then

$$J_2 = \sum_{i=1}^{n} E[\delta_{(0,0,x_3)}(\varepsilon w^1(1),\varepsilon w^2(1),2\varepsilon^2 S(1,w))\Phi_i(\varepsilon w)]$$
$$= \sum_{i=1}^{n}\int_{|\theta-\theta_i|<c(\eta)} E[\delta_{(0,0,x_3)}(\varepsilon w^1(1),\varepsilon w^2(1),2\varepsilon^2 S(1,w))\cdot$$
$$\delta_0((h_{\theta-(\pi/2)},\varepsilon w))\cdot(h_\theta,\varepsilon w)_H\Phi_i(\varepsilon w)]d\theta$$

and applying the Cameron-Martin theorem to the translation $w \longrightarrow w + h_\theta/\varepsilon$ yields that

$$J_2 = \sum_{i=1}^{n}\int_{|\theta-\theta_i|<c(\eta)} \exp\{-\|h_\theta\|_H^2/2\varepsilon^2\}E[\exp\{-(h_\theta,w)_H/\varepsilon\}\cdot$$
$$\delta_{(0,0,x_3)}(\varepsilon w^1(1),\varepsilon w^2(1),x_3+\frac{2\varepsilon}{\pi}(h_\theta,w)_H+2\varepsilon^2 S(1,w))\cdot$$

$$\delta_0((h_{\theta-(\pi/2)},h_\theta)_H+\varepsilon(h_{\theta-(\pi/2)},w)_H)(\|h_\theta\|_H^2+\varepsilon(h_\theta,w)_H)\cdot \Phi_i(\varepsilon w+h_\theta)]d\theta \ .$$

Noting that $(h_\theta,h_{\theta'})_H = 4\pi^2r^2\cos(\theta-\theta')$,

$$\begin{aligned}
J_2 &= \exp\{-\pi x_3/2\varepsilon^2\}\sum_{i=1}^{n}\int_{|\theta-\theta_i|<c(\eta)} E[\exp\{-(h_\theta,w)_H/\varepsilon\}\cdot \\
&\quad \delta_{(0,0,0,0)}(\varepsilon w^1(1),\varepsilon w^2(1),\tfrac{2\varepsilon}{\pi}[(h_\theta,w)_H+\varepsilon\pi S(1,w)], \\
&\quad \varepsilon(h_{\theta-(\pi/2)},w)_H)(4\pi^2r^2+\varepsilon(h_\theta,w)_H)\Phi_i(\varepsilon w+h_\theta)]d\theta \\
&= \varepsilon^{-4}2\pi^3r^2\exp\{-\pi x_3/2\varepsilon^2\}\sum_{i=1}^{n}\int_{|\theta-\theta_i|<c(\eta)} E[\exp\{\pi S(1,w)\}\cdot \\
&\quad \delta_{(0,0,0,0)}(w^1(1),w^2(1),(h_\theta,w)_H+\varepsilon\pi S(1,w),(h_{\theta-(\pi/2)},w)_H) \\
&\quad \times(1+\tfrac{\varepsilon}{4\pi^2r^2}(h_\theta,w)_H)\Phi_i(\varepsilon w+h_\theta)]d\theta \\
&= \varepsilon^{-4}2\pi^3r^2\exp\{-\pi x_3/2\varepsilon^2\}\sum_{i=1}^{n}\int_{|\theta-\theta_i|<c(\eta)} E[\exp\{\pi S(1,w)\}\cdot \\
&\quad \delta_{(0,0,0,0)}(g_{\varepsilon,\theta}(w))\chi(g_{\varepsilon,\theta}(w))\{1+\tfrac{\varepsilon}{4\pi^2r^2}(h_\theta,w)_H\}\cdot \\
&\quad \Phi_i(\varepsilon w+h_\theta)]d\theta \ .
\end{aligned}$$

Lemma 2.3 *In the above, we can choose* γ *and* η *such that, for each* $i = 1,2,\cdots,n$,

$$\begin{aligned}
(2.18)\qquad & \exp\{\pi S(1,w)\}\chi(g_{\varepsilon,\theta}(w))\Phi_i(\varepsilon w+h_\theta) \\
&= \exp\{\pi S(1,w)\}\chi(g_{0,\theta}(w))\Phi_i(h_\theta) + O(\varepsilon) \quad in\ \tilde{\mathbb{D}}^\infty \quad as\ \varepsilon\searrow 0 \\
&\qquad\qquad if\ |\theta-\theta_i| < c(\eta) \ .
\end{aligned}$$

Furthermore, $O(\varepsilon)$ *is uniform on* $\{\ \theta\ ;\ |\theta-\theta_i| < c(\eta)\ \}$.

Proof. Clearly, η can be made arbitrarily small accordingly as γ is made small in the above. Then, $\|\varepsilon w+h_\theta-h_{\theta_i}\|_2{}^2 < 2\gamma$ and $|\theta-\theta_i| < c(\eta)$ imply that

$$(2.19)\qquad \varepsilon^2\int_0^1|w(s)|^2ds \le 2\gamma + \text{const.}c(\eta) := \lambda$$

and λ can be made arbitrarily small. Using the same notation as in the proof of Lemma 2.1 ,

$$\pi S(1,w) = I_1 + I_2 + I_3 .$$

The condition $\chi(g_{\varepsilon,\theta}(w)) > 0$ implies that

$$|w^1(1)| + |w^2(1)| + |(h_\theta,w)_H+\varepsilon\pi S(1,w)| + |(h_{\theta-\pi/2},w)_H| < \delta$$

for some $\delta > 0$. We know that

$$E[e^{\alpha I_2}] < \infty \qquad \text{for any } |\alpha| < 2$$

and

$$E[e^{\alpha I_3};|w^1(1)|+|w^2(1)|<\delta] < \infty \qquad \text{for any } 1 < \alpha < \infty .$$

Furthermore, under the condition that

$$|(h_\theta,w)_H+\varepsilon\pi S(1,w)| + |(h_{\theta-(\pi/2)},w)_H| < \delta$$

I_1 can be estimated as

$$I_1 \le -(\xi_1^{(1)})^2 - (\xi_1^{(2)})^2 + c_1|\xi_1^{(1)}| + c_2|\xi_1^{(2)}|$$
$$+ c_3|\xi_1^{(1)}|\cdot|\varepsilon S(1,w)| + c_4|\xi_1^{(2)}|\cdot|\varepsilon S(1,w)|$$

where c_i are constants independent of ε . Hence, for any $\kappa > 0$,

$$(2.20) \quad I_1 \le -(\xi_1^{(1)})^2 - (\xi_1^{(2)})^2 + c_1|\xi_1^{(1)}| + c_2|\xi_1^{(2)}|$$
$$+ \frac{c_3}{2}(\kappa|\xi_1^{(1)}|^2+ \frac{1}{\kappa}|\varepsilon S(1,w)|^2) + \frac{c_4}{2}(\kappa|\xi_1^{(2)}|^2+ \frac{1}{\kappa}|\varepsilon S(1,w)|^2).$$

Since $\varepsilon S(1,w) = B\left(\varepsilon^2\int_0^1|w(s)|^2ds\right)$ for some 1-dimensional Brownian motion $B(t)$ with $B(0)=0$, we see easily that, for every $M > 1$, $\delta_1 > 0$ can be chosen such that

$$\max_{0<\varepsilon<1} E[\exp\{M(\varepsilon S(1,w))^2\};\varepsilon^2\int_0^1|w(s)|^2ds < \delta_1] < \infty .$$

Then, from (2.19) and (2.20) we easily deduce the following : For every $\alpha > 1$, we can choose γ and η in the above such that

$$\max_{|\theta-\theta_i|<c(\eta),0<\varepsilon<1} E[e^{\alpha I_1}\chi(g_{\varepsilon,\theta}(w)); \|\varepsilon w+h_\theta-h_{\theta_i}\|_2{}^2<2\gamma] < \infty .$$

Thus, we obtained that, for any $1<p<2$, we can choose γ and η such that

$$\max_{|\theta-\theta_i|<c(\eta),0<\varepsilon<1} E[e^{p\pi S(1,w)}\chi(g_{\varepsilon,\theta}(w))\Phi_i(\varepsilon w+h_\theta)] < \infty ,$$
$$i = 1,2,\cdots,n .$$

(2.18) is easily concluded from this (cf.[12]) . //

By a general result in [12], we know that

$$(2.21) \qquad \delta_{(0,0,0,0)}(g_{\varepsilon,\theta}(w)) = \delta_{(0,0,0,0)}(g_{0,\theta}(w)) + O(\varepsilon)$$
$$in \ \tilde{\mathbb{D}}^{-\infty} \ as \ \varepsilon \searrow 0 .$$

This, combined with (2.18), yields that

$$J_2 = \varepsilon^{-4}2\pi^3 r^2\exp\{-\pi x_3/2\varepsilon^2\}\Big[\sum_{i=1}^n\int_{|\theta-\theta_i|<c(\eta)} E[\exp\{\pi S(1,w)\}\cdot$$
$$\delta_{(0,0,0,0)}(g_{0,\theta}(w))\chi(g_{0,\theta}(w))]\Phi_i(h_\theta)d\theta + O(\varepsilon)\Big] .$$

Hence by (2.13)

$$J_2 \sim \varepsilon^{-4}\frac{1}{8\pi}\exp\{-\pi x_3/2\varepsilon^2\}\sum_{i=1}^n\int_{|\theta-\theta_i|<c(\eta)}\Phi_i(h_\theta)d\theta .$$

Finally we have

$$\sum_{i=1}^n\int_{|\theta-\theta_i|<c(\eta)}\Phi_i(h_\theta)d\theta = \sum_{i=1}^n\int_{|\theta-\theta_i|<c(\eta)} I_{U_i}(h_\theta)\Phi_i(h_\theta)d\theta$$
$$= \sum_{i=1}^n\int_0^{2\pi} I_{U_i}(h_\theta)\Phi_i(h_\theta)d\theta = \sum_{i=1}^n\int_0^{2\pi}\Phi_i(h_\theta)d\theta = 2\pi$$

because $h_\theta \in U_i$ implies $|\theta-\theta_i| < c(\eta)$ and $\sum_{i=1}^n \Phi_i(h_\theta) = 1$, $\theta \in [0,2\pi)$. Thus we obtained

$$J_2 \sim \frac{1}{4\varepsilon^4}\exp\{-\pi x_3/2\varepsilon^2\}$$

and this, combined with (2.16), completes the proof of (2.10).

§ 3 The case of the group $N_{4,2}$.

As we saw in Section 1, the group $N_{4,2}$ is realized by $\mathbb{R}^{10}$ and $x \in \mathbb{R}^{10}$ is denoted by $x = (x_i,x_{ij})_{1\le i<j\le 4}$. We identify (x_{ij})

with the 4×4 skew symmetric matrix $X = (X_{ij})$ defined by

$$X_{ij} = \begin{cases} x_{ij} , & i < j \\ -x_{ji} , & i > j \\ 0 , & i = j \end{cases}$$

and write $x = [\,\underline{x}, X\,]$ where $\underline{x} = (\,x_1, x_2, x_3, x_4\,)$. For $T \in O(4)$, define an action of T on $\mathbb{R}^{10}$ by

$$Tx = [\,T\underline{x}, TX{}^tT\,] .$$

We consider the vector fields V_i , $i=1,2,3,4$ on $\mathbb{R}^{10}$ defined by (1.16). It is easy to see that (H.1) is satisfied everywhere although (H.2) is violated at every point. Hence the fundamental solution exists. Since it is invariant under the group action, we may consider $p(t,0,x)$, $x \in \mathbb{R}^{10}$, only. In this note, we restrict ourselves to the case of $x = (\,\underline{0}\,,\, x_{ij}\,)$. Since p is also invariant under the action of $O(4)$, we may assume $x = [\,\underline{0}, V\,]$ where V is of the form

$$V = \begin{pmatrix} 0 & v_1 & & \\ -v_1 & 0 & & 0 \\ & 0 & 0 & v_2 \\ & & -v_2 & 0 \end{pmatrix} , \qquad 0 \le v_2 \le v_1 ,$$

that is, $x = (\,\underline{0}\,,\, v_1\delta_{1i}\delta_{2j} + v_2\delta_{3i}\delta_{4j}\,)$. Let (W^4, P) be the 4-dimensional Wiener space. L-diffusion $X(t)$ starting at $0 \in \mathbb{R}^{10}$ is realized on this space by

$$(3.1) \qquad \begin{cases} x_i(t,w) = w^i(t) , & i=1,2,3,4 \\ x_{ij}(t,w) = S^{ij}(t,w) , & 1 \le i < j \le 4 \end{cases}$$

where

$$(3.2) \qquad S^{ij}(t,w) = \frac{1}{2}\Big[\int_0^t w^i(s)dw^j(s) - w^j(s)dw^i(s)\Big] .$$

Let $H^4 \subset W^4$ be the Cameron-Martin subspace. For $\mathbf{h} = (h^1, h^2, h^3, h^4) \in H^4$, set $\mathbf{h}^{[1]} = (h^1, h^2)$, $\mathbf{h}^{[2]} = (h^3, h^4)$ and write $\mathbf{h} =$

$(h^{[1]},h^{[2]})$. Note that $h^{[i]} \in H^2$, $i=1,2$, H^2 being the 2-dimensional Cameron-Martin subspace (denoted by H in section 2). Clearly $p(\varepsilon^2,0,0) = \varepsilon^{-16}E[\delta_{(0,\cdots,0)}(w(1),S^{ij}(1,w))]$. So we omit the case $x=0$, i.e., we assume $v_1>0$.

Case I , $v_2 = 0$ and $v := v_1 > 0$.

In this case, $K^{0,x}_{\min} = \{ h_\theta = (h^{[1]}_\theta, h^{[2]}_\theta) , \theta \in [0,2\pi) \}$ where $h^{[1]}_\theta = h_{\theta,\sqrt{v/\pi}}$, and $h^{[2]}_\theta = 0$. Here $h_{\theta,r} = (h^1_{\theta,r}, h^2_{\theta,r}) \in H^2$, $\theta \in [0,2\pi)$, $r>0$, is defined by

$$\text{(3.3)} \qquad \begin{cases} h^1_{\theta,r}(t) = r\cos\theta\cdot\sin 2\pi t - r\sin\theta(1-\cos 2\pi t) \\ h^2_{\theta,r}(t) = r\cos\theta(1-\cos 2\pi t) + r\sin\theta\cdot\sin 2\pi t \end{cases} .$$

We have

$$p(\varepsilon^2,0,x) = E[\delta_{(\underline{0},v\delta_{1i}\delta_{2j})}(\varepsilon w(1),\varepsilon^2 S(1,w))]$$

and we do the same analysis as in Section 2 (e.g. the Cameron-Martin theorem and large deviation argument). Then we have,

$$\text{(3.4)} \qquad p(\varepsilon^2,0,x) \sim 16\pi^3 v\cdot\varepsilon^{-12}\exp\{-\frac{2\pi v}{\varepsilon^2}\}\cdot E[\exp\{2\pi S^{12}(1,w)\}\cdot\delta_{\underline{0}}(w(1)) \cdot\delta_{(0,0)}((h_{\theta,\sqrt{v/\pi}},w^{[1]})_{H^2},(h_{\theta-(\pi/2),\sqrt{v/\pi}},w^{[1]})_{H^2})\cdot \prod_{\substack{i=1,2\\ j=3,4}}\delta_0(\int_0^1 h^i_{\theta,\sqrt{v/\pi}}(s)dw^j(s))\delta_0(S^{34}(1,w))] \quad \text{as } \varepsilon\searrow 0$$

where we use the notation $w^{[1]}(t) = (w^1(t),w^2(t))$ and $w^{[2]} = (w^3(t),w^4(t))$. Since $\exp\{2\pi S^{12}(1,w)\}\cdot\chi(w^{[1]}(1),(h_{\theta,\sqrt{v/\pi}},w^{[1]})_{H^2}, (h_{\theta-(\pi/2),\sqrt{v/\pi}},w^{[1]})_{H^2}) \in \mathbb{D}^\infty$ by Lemma 2.1, the above generalized expectation is well-defined. For the non-degeneracy of 11-dimensional Wiener functional appearing in δ-functions, cf. [10].

Case II , $0 < v_2 < v_1$.

In this case, $K^{0,x}_{\min} = \{ h_{\theta_1,\theta_2} = (h_{\theta_1,\sqrt{v_1/\pi}} , h_{\theta_2,\sqrt{v_2/2\pi}}) ,$

$(\theta_1,\theta_2) \in [0,2\pi)^2 \}$. Thus $K^{0,x}_{\min}$ is a two dimensional torus. We can apply the same method as in Section 2 to obtain

$$(3.5) \quad p(\varepsilon^2,0,x) \sim \varepsilon^{-12}2^7\pi^4 v_1 v_2 \exp\{-(4\pi v_1+8\pi v_2)/2\varepsilon^2\}$$
$$\times E[\exp\{2\pi S^{12}(1,w)+4\pi S^{34}(1,w)\}\delta_{\underline{0}}(w(1))$$
$$\prod_{1\le i<j\le 4}\delta_0(\int_0^1 h^i_{\theta_1,\theta_2}(t)dw^j(t)-\int_0^1 h^j_{\theta_1,\theta_2}(t)dw^i(t))$$
$$\prod_{k=1}^{2}\delta_0((h_{\theta_k-(\pi/2),\sqrt{v_k/k\pi}}, w^{[k]})_{H^2})] .$$

Here we only note that the above generalized expectation is well-defined and is independent of (θ_1,θ_2) . For this, we introduce the following notations :

$$\Xi^i = w^i(1) , \qquad i=1,2,3,4$$
$$\Xi^{ij} = \int_0^1 h^i_{\theta_1,\theta_2}(t)dw^j(t) - \int_0^1 h^j_{\theta_1,\theta_2}(t)dw^i(t) , \quad 1\le i<j\le 4$$

and

$$\Xi_k = (h_{\theta_k-(\pi/2),\sqrt{v_k/k\pi}} , w^{[k]})_{H^2} , \qquad k=1,2 .$$

Also we set

$$\xi^{(i)}_m = -\sqrt{2}\int_0^1 \sin 2\pi mt\,dw^i(t) , \qquad \eta^{(i)}_m = \sqrt{2}\int_0^1 \cos 2\pi mt\,dw^i(t) ,$$
$$i=1,2,3,4,\ m=1,2,\cdots .$$

Then

$$S^{ij}(1,w) = \frac{1}{2\pi}\sum_{m=1}^{\infty}\frac{1}{m}\{ \xi^{(i)}_m(\eta^{(j)}_m-\sqrt{2}\cdot\Xi^j) - \xi^{(j)}_m(\eta^{(i)}_m-\sqrt{2}\cdot\Xi^i) \} ,$$
$$\Xi^{12} = \sqrt{2}\ (-\sqrt{v_1/4\pi}\cdot\cos\theta_1\cdot\xi^{(2)}_1 + \sqrt{v_1/4\pi}\cdot\sin\theta_1\cdot\eta^{(2)}_1$$
$$+ \sqrt{v_1/4\pi}\cdot\cos\theta_1\cdot\eta^{(1)}_1 + \sqrt{v_1/4\pi}\cdot\sin\theta_1\cdot\xi^{(1)}_1)$$
$$\Xi^{13} = \sqrt{2}\ (-\sqrt{v_1/4\pi}\cdot\cos\theta_1\cdot\xi^{(3)}_1 + \sqrt{v_1/4\pi}\cdot\sin\theta_1\cdot\eta^{(3)}_1$$
$$+ \sqrt{v_2/8\pi}\cdot\cos\theta_2\cdot\xi^{(1)}_2 - \sqrt{v_2/8\pi}\cdot\sin\theta_2\cdot\eta^{(1)}_2)$$
$$\Xi^{14} = \sqrt{2}\ (-\sqrt{v_1/4\pi}\cdot\cos\theta_1\cdot\xi^{(4)}_1 + \sqrt{v_1/4\pi}\cdot\sin\theta_1\cdot\eta^{(4)}_1$$
$$+ \sqrt{v_2/8\pi}\cdot\cos\theta_2\cdot\eta^{(1)}_2 + \sqrt{v_2/8\pi}\cdot\sin\theta_2\cdot\xi^{(1)}_2)$$
$$\Xi^{23} = \sqrt{2}\ (\sqrt{v_2/8\pi}\cdot\cos\theta_2\cdot\xi^{(2)}_2 - \sqrt{v_2/8\pi}\cdot\sin\theta_2\cdot\eta^{(2)}_2$$
$$- \sqrt{v_1/4\pi}\cdot\cos\theta_1\cdot\eta^{(3)}_1 - \sqrt{v_1/4\pi}\cdot\sin\theta_1\cdot\xi^{(3)}_1)$$

$$\Xi^{24} = \sqrt{2}\ (-\sqrt{v_1/4\pi}\cdot\cos\theta_1\cdot\eta_1^{(4)} - \sqrt{v_1/4\pi}\cdot\sin\theta_1\cdot\xi_1^{(4)} + \sqrt{v_2/8\pi}\cdot\cos\theta_2\cdot\eta_2^{(2)} + \sqrt{v_2/8\pi}\cdot\sin\theta_2\cdot\xi_2^{(2)})$$

$$\Xi^{34} = \sqrt{2}\ (-\sqrt{v_2/8\pi}\cdot\cos\theta_2\cdot\xi_2^{(4)} + \sqrt{v_2/8\pi}\cdot\sin\theta_2\cdot\eta_2^{(4)} + \sqrt{v_2/8\pi}\cdot\cos\theta_2\cdot\eta_2^{(3)} + \sqrt{v_2/8\pi}\cdot\sin\theta_2\cdot\xi_2^{(3)})$$

and

$$\Xi_k = \sqrt{2k\pi v_k}\ (\sin\theta_k\cdot\eta_k^{(2k-1)} - \cos\theta_k\cdot\xi_k^{(2k-1)} - \sin\theta_k\cdot\xi_k^{(2k)} - \cos\theta_k\cdot\eta_k^{(2k)})\ ,\quad k=1,2\ .$$

It is easy to check that

$$\bar{\Xi} = (\Xi^1,\cdots,\Xi^4,\Xi^{12},\cdots,\Xi^{34},\Xi_1,\Xi_2)$$

is a non-degenerate 12-dimensional Gaussian random variable whose law is independent of (θ_1,θ_2) . The generalized expectation appearing in the right-hand side of (3.5) is $(2\pi)^{-6}\det(\mathrm{cov}(\bar{\Xi}))^{-1/2}\cdot I$ where

$$\begin{aligned} I &= \mathrm{E}[\exp(\sum_{k=1}^{2} \sum_{m=1}^{\infty} \frac{k}{m} (\xi_m^{(2k-1)}(\eta_m^{(2k)}-\sqrt{2}\cdot\Xi^{2k})- \\ &\qquad \xi_m^{(2k)}(\eta_m^{(2k-1)}-\sqrt{2}\cdot\Xi^{2k-1})))|\bar{\Xi}=0] \\ &= \prod_{k=1}^{2} \mathrm{E}[\exp(\xi_k^{(2k-1)}\eta_k^{(2k)}-\xi_k^{(2k)}\eta_k^{(2k-1)})|\ \Xi^{2k-1,2k}=0,\ \Xi_k=0] \\ &\quad\times \mathrm{E}[\exp\{\tfrac{1}{2}(\xi_2^{(1)}\eta_2^{(2)}-\xi_2^{(2)}\eta_2^{(1)}) + 2(\xi_1^{(3)}\eta_1^{(4)}-\xi_1^{(4)}\eta_1^{(3)})\}| \\ &\qquad \Xi^{13} = \Xi^{14} = \Xi^{23} = \Xi^{24} = 0\ ] \\ &\quad\times \prod_{k=1}^{2} \prod_{m=3}^{\infty} \mathrm{E}[\exp\{ \frac{k}{m} (\xi_m^{(2k-1)}\eta_m^{(2k)} - \xi_m^{(2k)}\eta_m^{(2k-1)})\}\ ] \\ &:= I_1 \times I_2 \times I_3\ . \end{aligned}$$

In the same way as in the proof of Lemma 2.1, $I_1 < \infty$. Also, $I_3 = \prod_{k=1}^{2} \prod_{m=3}^{\infty} \left(1 - \frac{k^2}{m^2}\right)^{-1} < \infty$. It is easy to see that $\Xi^{13} = \Xi^{14} = \Xi^{23} = \Xi^{24} = 0$ if and only if $\bar{\xi} = P\bar{\eta}$ where

$$P = \frac{1}{q}\begin{pmatrix} p_1p_3 & -p_1p_4 & p_2p_3 & -p_2p_4 \\ -p_2p_3 & p_2p_4 & p_1p_3 & -p_1p_4 \\ p_1p_4 & p_1p_3 & p_2p_4 & p_2p_3 \\ -p_2p_4 & -p_2p_3 & p_1p_4 & p_1p_3 \end{pmatrix},$$

$$\bar{\xi} = {}^t(\xi_1^{(3)},\eta_1^{(3)},\xi_1^{(4)},\eta_1^{(4)})\ ,\quad \bar{\eta} = {}^t(\xi_2^{(1)},\eta_2^{(1)},\xi_2^{(2)},\eta_2^{(2)})\ ,$$

with $p_1 = \sqrt{4\pi v_1}\cos\theta_1$, $p_2 = \sqrt{4\pi v_1}\sin\theta_1$, $p_3 = \sqrt{8\pi v_2}\cos\theta_2$, $p_4 = \sqrt{8\pi v_2}\sin\theta_2$ and $q = 8\pi v_1$. Note that $\sqrt{2v_1/v_2}\cdot P \in O(4)$. Then, under the condition that $\bar{\xi} - P\bar{\eta} = 0$,

$$\frac{1}{2}\left(\xi_2^{(1)}\eta_2^{(2)} - \xi_2^{(2)}\eta_2^{(1)} \right) + 2\left(\xi_1^{(3)}\eta_1^{(4)} - \xi_1^{(4)}\eta_1^{(3)} \right)$$
$$= \left(\frac{1}{2} + (v_2/v_1) \right)\left(\xi_2^{(1)}\eta_2^{(2)} - \xi_2^{(2)}\eta_2^{(1)} \right)$$

and it is easy to see that

$$I_2 = [\frac{1}{2\pi}(1+(v_2/2v_1))]^2 \int_{\mathbb{R}^4} \exp[(\frac{1}{2}+(v_2/v_1))(x_1x_4-x_2x_3)]\cdot$$
$$\cdot\exp(-\frac{1}{2}(1+(v_2/2v_1))|x|^2)dx < \infty \quad \text{if} \quad v_1 > v_2 .$$

Case III , $\quad 0 < v_1 = v_2 := v$.

In this case, $K^{0,x}_{\min} \subset H^4$ is a 4-dimensional compact manifold and, by the same method, we can obtain

(3.6) $\quad p(\varepsilon^2,0,x) \sim \text{const.}\varepsilon^{-14}\exp(-6\pi v/\varepsilon^2)$.

Details will be discussed elsewhere.

References.

[1] R.Azencott : Diffusions invariantes sur le groupes d'Heisenberg ; une étude de cas d'après B.Gaveau , *Géodésiques et diffusions en temps petit* , Astérisque 84-85 (1981) , 227-235 .

[2] R.Azencott : Densités des diffusions en temps petit : développements asymptotiques . Séminaire de Prob. XVIII , 1982/1983 Lecture Note in Math. **1059** , Springer (1984) , 402-498 .

[3] G.Ben Arous : Développement asymptotique du noyau de la chaleur hypoelliptique hors du cut-locus , preprint .

[4] J.-M. Bismut : *Large deviations and the Malliavin calculus* , Progress in Math. **45** , Birkhäuser , 1984 .

[5] B.Gaveau : Principe de moindre action, propagation de la chaleur, estimeés sous elliptiques sur certains groupes nilpotents Acta Math. **139** (1977) , 95-153 .

[6] I.M.Gelfand and G.E.Silov : *Generalized functions* , Vol.1 , Academic Press , 1964 .

[7] S.Kusuoka and D.W.Stroock : Applications of the Malliavin calculus , Part III J. Fac. Sci. Univ. Tokyo Sect. IA Math. 34(1987), 391-442 .

[8] R.Léandre : Intégration dans la fibre associée à une diffusion dégénérée , Probab. Th. Rel. Fields **76** (1987) , 341-358

[9] S.A.Molchanov : Diffusion processes and Riemannian geometry , Russian Math. Surveys **30** (1975) , 1-63 .

[10] S.Takanobu : Diagonal short time asymptotics of heat kernels for certain degenerate second order differential operators of Hörmander type , to appear in Publ. RIMS , Kyoto Univ. .

[11] H.Uemura : On a short time expansion of the fundamental solution of heat equations by the method of Wiener functionals , J. Math. Kyoto Univ. **27** (1987) , 417-431 .

[12] S.Watanabe : Analysis of Wiener functionals (Malliavin calculus) and its applications to heat kernels , The Annals of Probab. **15** (1987) , 1-39 .

Vol. 1145: G. Winkler, Choquet Order and Simplices. VI, 143 pages. 1985.

Vol. 1146: Séminaire d'Algèbre Paul Dubreil et Marie-Paule Malliavin. Proceedings, 1983–1984. Edité par M.-P. Malliavin. IV, 420 pages. 1985.

Vol. 1147: M. Wschebor, Surfaces Aléatoires. VII, 111 pages. 1985.

Vol. 1148: Mark A. Kon, Probability Distributions in Quantum Statistical Mechanics. V, 121 pages. 1985.

Vol. 1149: Universal Algebra and Lattice Theory. Proceedings, 1984. Edited by S. D. Comer. VI, 282 pages. 1985.

Vol. 1150: B. Kawohl, Rearrangements and Convexity of Level Sets in PDE. V, 136 pages. 1985.

Vol 1151: Ordinary and Partial Differential Equations. Proceedings, 1984. Edited by B.D. Sleeman and R.J. Jarvis. XIV, 357 pages. 1985.

Vol. 1152: H. Widom, Asymptotic Expansions for Pseudodifferential Operators on Bounded Domains. V, 150 pages. 1985.

Vol. 1153: Probability in Banach Spaces V. Proceedings, 1984. Edited by A. Beck, R. Dudley, M. Hahn, J. Kuelbs and M. Marcus. VI, 457 pages. 1985.

Vol. 1154: D.S. Naidu, A.K. Rao, Singular Pertubation Analysis of Discrete Control Systems. IX, 195 pages. 1985.

Vol. 1155: Stability Problems for Stochastic Models. Proceedings, 1984. Edited by V.V. Kalashnikov and V.M. Zolotarev. VI, 447 pages. 1985.

Vol. 1156: Global Differential Geometry and Global Analysis 1984. Proceedings, 1984. Edited by D. Ferus, R.B. Gardner, S. Helgason and U. Simon. V, 339 pages. 1985.

Vol. 1157: H. Levine, Classifying Immersions into $\mathbb{R}^4$ over Stable Maps of 3-Manifolds into $\mathbb{R}^2$. V, 163 pages. 1985.

Vol. 1158: Stochastic Processes – Mathematics and Physics. Proceedings, 1984. Edited by S. Albeverio, Ph. Blanchard and L. Streit. VI, 230 pages. 1986.

Vol. 1159: Schrödinger Operators, Como 1984. Seminar. Edited by S. Graffi. VIII, 272 pages. 1986.

Vol. 1160: J.-C. van der Meer, The Hamiltonian Hopf Bifurcation. VI, 115 pages. 1985.

Vol. 1161: Harmonic Mappings and Minimal Immersions, Montecatini 1984. Seminar. Edited by E. Giusti. VII, 285 pages. 1985.

Vol. 1162: S.J.L. van Eijndhoven, J. de Graaf, Trajectory Spaces, Generalized Functions and Unbounded Operators. IV, 272 pages. 1985.

Vol. 1163: Iteration Theory and its Functional Equations. Proceedings, 1984. Edited by R. Liedl, L. Reich and Gy. Targonski. VIII, 231 pages. 1985.

Vol. 1164: M. Meschiari, J.H. Rawnsley, S. Salamon, Geometry Seminar "Luigi Bianchi" II – 1984. Edited by E. Vesentini. VI, 224 pages. 1985.

Vol. 1165: Seminar on Deformations. Proceedings, 1982/84. Edited by J. Ławrynowicz. IX, 331 pages. 1985.

Vol. 1166: Banach Spaces. Proceedings, 1984. Edited by N. Kalton and E. Saab. VI, 199 pages. 1985.

Vol. 1167: Geometry and Topology. Proceedings, 1983–84. Edited by J. Alexander and J. Harer. VI, 292 pages. 1985.

Vol. 1168: S.S. Agaian, Hadamard Matrices and their Applications. III, 227 pages. 1985.

Vol. 1169: W.A. Light, E.W. Cheney, Approximation Theory in Tensor Product Spaces. VII, 157 pages. 1985.

Vol. 1170: B.S. Thomson, Real Functions. VII, 229 pages. 1985.

Vol. 1171: Polynômes Orthogonaux et Applications. Proceedings, 1984. Edité par C. Brezinski, A. Draux, A.P. Magnus, P. Maroni et A. Ronveaux. XXXVII, 584 pages. 1985.

Vol. 1172: Algebraic Topology, Göttingen 1984. Proceedings. Edited by L. Smith. VI, 209 pages. 1985.

Vol. 1173: H. Delfs, M. Knebusch, Locally Semialgebraic Spaces. XVI, 329 pages. 1985.

Vol. 1174: Categories in Continuum Physics, Buffalo 1982. Seminar. Edited by F.W. Lawvere and S.H. Schanuel. V, 126 pages. 1986.

Vol. 1175: K. Mathiak, Valuations of Skew Fields and Projective Hjelmslev Spaces. VII, 116 pages. 1986.

Vol. 1176: R.R. Bruner, J.P. May, J.E. McClure, M. Steinberger, H_∞ Ring Spectra and their Applications. VII, 388 pages. 1986.

Vol. 1177: Representation Theory I. Finite Dimensional Algebras. Proceedings, 1984. Edited by V. Dlab, P. Gabriel and G. Michler. XV, 340 pages. 1986.

Vol. 1178: Representation Theory II. Groups and Orders. Proceedings, 1984. Edited by V. Dlab, P. Gabriel and G. Michler. XV, 370 pages. 1986.

Vol. 1179: Shi J.-Y. The Kazhdan-Lusztig Cells in Certain Affine Weyl Groups. X, 307 pages. 1986.

Vol. 1180: R. Carmona, H. Kesten, J.B. Walsh, École d'Été de Probabilités de Saint-Flour XIV – 1984. Édité par P.L. Hennequin. X, 438 pages. 1986.

Vol. 1181: Buildings and the Geometry of Diagrams, Como 1984. Seminar. Edited by L. Rosati. VII, 277 pages. 1986.

Vol. 1182: S. Shelah, Around Classification Theory of Models. VII, 279 pages. 1986.

Vol. 1183: Algebra, Algebraic Topology and their Interactions. Proceedings, 1983. Edited by J.-E. Roos. XI, 396 pages. 1986.

Vol. 1184: W. Arendt, A. Grabosch, G. Greiner, U. Groh, H.P. Lotz, U. Moustakas, R. Nagel, F. Neubrander, U. Schlotterbeck, One-parameter Semigroups of Positive Operators. Edited by R. Nagel. X, 460 pages. 1986.

Vol. 1185: Group Theory, Beijing 1984. Proceedings. Edited by Tuan H.F. V, 403 pages. 1986.

Vol. 1186: Lyapunov Exponents. Proceedings, 1984. Edited by L. Arnold and V. Wihstutz. VI, 374 pages. 1986.

Vol. 1187: Y. Diers, Categories of Boolean Sheaves of Simple Algebras. VI, 168 pages. 1986.

Vol. 1188: Fonctions de Plusieurs Variables Complexes V. Séminaire, 1979–85. Edité par François Norguet. VI, 306 pages. 1986.

Vol. 1189: J. Lukeš, J. Malý, L. Zajíček, Fine Topology Methods in Real Analysis and Potential Theory. X, 472 pages. 1986.

Vol. 1190: Optimization and Related Fields. Proceedings, 1984. Edited by R. Conti, E. De Giorgi and F. Giannessi. VIII, 419 pages. 1986.

Vol. 1191: A.R. Its, V.Yu. Novokshenov, The Isomonodromic Deformation Method in the Theory of Painlevé Equations. IV, 313 pages. 1986.

Vol. 1192: Equadiff 6. Proceedings, 1985. Edited by J. Vosmansky and M. Zlámal. XXIII, 404 pages. 1986.

Vol. 1193: Geometrical and Statistical Aspects of Probability in Banach Spaces. Proceedings, 1985. Edited by X. Femique, B. Heinkel, M.B. Marcus and P.A. Meyer. IV, 128 pages. 1986.

Vol. 1194: Complex Analysis and Algebraic Geometry. Proceedings, 1985. Edited by H. Grauert. VI, 235 pages. 1986.

Vol. 1195: J.M. Barbosa, A.G. Colares, Minimal Surfaces in $\mathbb{R}^3$. X, 124 pages. 1986.

Vol. 1196: E. Casas-Alvero, S. Xambó-Descamps, The Enumerative Theory of Conics after Halphen. IX, 130 pages. 1986.

Vol. 1197: Ring Theory. Proceedings, 1985. Edited by F.M.J. van Oystaeyen. V, 231 pages. 1986.

Vol. 1198: Séminaire d'Analyse, P. Lelong – P. Dolbeault – H. Skoda. Seminar 1983/84. X, 260 pages. 1986.

Vol. 1199: Analytic Theory of Continued Fractions II. Proceedings, 1985. Edited by W.J. Thron. VI, 299 pages. 1986.

Vol. 1200: V.D. Milman, G. Schechtman, Asymptotic Theory of Finite Dimensional Normed Spaces. With an Appendix by M. Gromov. VIII, 156 pages. 1986.

Vol. 1201: Curvature and Topology of Riemannian Manifolds. Proceedings, 1985. Edited by K. Shiohama, T. Sakai and T. Sunada. VII, 336 pages. 1986.

Vol. 1202: A. Dür, Möbius Functions, Incidence Algebras and Power Series Representations. XI, 134 pages. 1986.

Vol. 1203: Stochastic Processes and Their Applications. Proceedings, 1985. Edited by K. Itô and T. Hida. VI, 222 pages. 1986.

Vol. 1204: Séminaire de Probabilités XX, 1984/85. Proceedings. Edité par J. Azéma et M. Yor. V, 639 pages. 1986.

Vol. 1205: B.Z. Moroz, Analytic Arithmetic in Algebraic Number Fields. VII, 177 pages. 1986.

Vol. 1206: Probability and Analysis, Varenna (Como) 1985. Seminar. Edited by G. Letta and M. Pratelli. VIII, 280 pages. 1986.

Vol. 1207: P.H. Bérard, Spectral Geometry: Direct and Inverse Problems. With an Appendix by G. Besson. XIII, 272 pages. 1986.

Vol. 1208: S. Kaijser, J.W. Pelletier, Interpolation Functors and Duality. IV, 167 pages. 1986.

Vol. 1209: Differential Geometry, Peñíscola 1985. Proceedings. Edited by A.M. Naveira, A. Ferrández and F. Mascaró. VIII, 306 pages. 1986.

Vol. 1210: Probability Measures on Groups VIII. Proceedings, 1985. Edited by H. Heyer. X, 386 pages. 1986.

Vol. 1211: M.B. Sevryuk, Reversible Systems. V, 319 pages. 1986.

Vol. 1212: Stochastic Spatial Processes. Proceedings, 1984. Edited by P. Tautu. VIII, 311 pages. 1986.

Vol. 1213: L.G. Lewis, Jr., J.P. May, M. Steinberger, Equivariant Stable Homotopy Theory. IX, 538 pages. 1986.

Vol. 1214: Global Analysis – Studies and Applications II. Edited by Yu.G. Borisovich and Yu.E. Gliklikh. V, 275 pages. 1986.

Vol. 1215: Lectures in Probability and Statistics. Edited by G. del Pino and R. Rebolledo. V, 491 pages. 1986.

Vol. 1216: J. Kogan, Bifurcation of Extremals in Optimal Control. VIII, 106 pages. 1986.

Vol. 1217: Transformation Groups. Proceedings, 1985. Edited by S. Jackowski and K. Pawalowski. X, 396 pages. 1986.

Vol. 1218: Schrödinger Operators, Aarhus 1985. Seminar. Edited by E. Balslev. V, 222 pages. 1986.

Vol. 1219: R. Weissauer, Stabile Modulformen und Eisensteinreihen. III, 147 Seiten. 1986.

Vol. 1220: Séminaire d'Algèbre Paul Dubreil et Marie-Paule Malliavin. Proceedings, 1985. Edité par M.-P. Malliavin. IV, 200 pages. 1986.

Vol. 1221: Probability and Banach Spaces. Proceedings, 1985. Edited by J. Bastero and M. San Miguel. XI, 222 pages. 1986.

Vol. 1222: A. Katok, J.-M. Strelcyn, with the collaboration of F. Ledrappier and F. Przytycki, Invariant Manifolds, Entropy and Billiards; Smooth Maps with Singularities. VIII, 283 pages. 1986.

Vol. 1223: Differential Equations in Banach Spaces. Proceedings, 1985. Edited by A. Favini and E. Obrecht. VIII, 299 pages. 1986.

Vol. 1224: Nonlinear Diffusion Problems, Montecatini Terme 1985. Seminar. Edited by A. Fasano and M. Primicerio. VIII, 188 pages. 1986.

Vol. 1225: Inverse Problems, Montecatini Terme 1986. Seminar. Edited by G. Talenti. VIII, 204 pages. 1986.

Vol. 1226: A. Buium, Differential Function Fields and Moduli of Algebraic Varieties. IX, 146 pages. 1986.

Vol. 1227: H. Helson, The Spectral Theorem. VI, 104 pages. 1986.

Vol. 1228: Multigrid Methods II. Proceedings, 1985. Edited by W. Hackbusch and U. Trottenberg. VI, 336 pages. 1986.

Vol. 1229: O. Bratteli, Derivations, Dissipations and Group Actions on C*-algebras. IV, 277 pages. 1986.

Vol. 1230: Numerical Analysis. Proceedings, 1984. Edited by J.-P. Hennart. X, 234 pages. 1986.

Vol. 1231: E.-U. Gekeler, Drinfeld Modular Curves. XIV, 107 pages. 1986.

Vol. 1232: P.C. Schuur, Asymptotic Analysis of Soliton Problems. VIII, 180 pages. 1986.

Vol. 1233: Stability Problems for Stochastic Models. Proceedings, 1985. Edited by V.V. Kalashnikov, B. Penkov and V.M. Zolotarev. VI, 223 pages. 1986.

Vol. 1234: Combinatoire énumérative. Proceedings, 1985. Edité par G. Labelle et P. Leroux. XIV, 387 pages. 1986.

Vol. 1235: Séminaire de Théorie du Potentiel, Paris, No. 8. Directeurs: M. Brelot, G. Choquet et J. Deny. Rédacteurs: F. Hirsch et G. Mokobodzki. III, 209 pages. 1987.

Vol. 1236: Stochastic Partial Differential Equations and Applications. Proceedings, 1985. Edited by G. Da Prato and L. Tubaro. V, 257 pages. 1987.

Vol. 1237: Rational Approximation and its Applications in Mathematics and Physics. Proceedings, 1985. Edited by J. Gilewicz, M. Pindor and W. Siemaszko. XII, 350 pages. 1987.

Vol. 1238: M. Holz, K.-P. Podewski and K. Steffens, Injective Choice Functions. VI, 183 pages. 1987.

Vol. 1239: P. Vojta, Diophantine Approximations and Value Distribution Theory. X, 132 pages. 1987.

Vol. 1240: Number Theory, New York 1984–85. Seminar. Edited by D.V. Chudnovsky, G.V. Chudnovsky, H. Cohn and M.B. Nathanson. V, 324 pages. 1987.

Vol. 1241: L. Gårding, Singularities in Linear Wave Propagation. III, 125 pages. 1987.

Vol. 1242: Functional Analysis II, with Contributions by J. Hoffmann-Jørgensen et al. Edited by S. Kurepa, H. Kraljević and D. Butković. VII, 432 pages. 1987.

Vol. 1243: Non Commutative Harmonic Analysis and Lie Groups. Proceedings, 1985. Edited by J. Carmona, P. Delorme and M. Vergne. V, 309 pages. 1987.

Vol. 1244: W. Müller, Manifolds with Cusps of Rank One. XI, 158 pages. 1987.

Vol. 1245: S. Rallis, L-Functions and the Oscillator Representation. XVI, 239 pages. 1987.

Vol. 1246: Hodge Theory. Proceedings, 1985. Edited by E. Cattani, F. Guillén, A. Kaplan and F. Puerta. VII, 175 pages. 1987.

Vol. 1247: Séminaire de Probabilités XXI. Proceedings. Edité par J. Azéma, P.A. Meyer et M. Yor. IV, 579 pages. 1987.

Vol. 1248: Nonlinear Semigroups, Partial Differential Equations and Attractors. Proceedings, 1985. Edited by T.L. Gill and W.W. Zachary. IX, 185 pages. 1987.

Vol. 1249: I. van den Berg, Nonstandard Asymptotic Analysis. IX, 187 pages. 1987.

Vol. 1250: Stochastic Processes – Mathematics and Physics II. Proceedings 1985. Edited by S. Albeverio, Ph. Blanchard and L. Streit. VI, 359 pages. 1987.

Vol. 1251: Differential Geometric Methods in Mathematical Physics. Proceedings, 1985. Edited by P.L. García and A. Pérez-Rendón. VII, 300 pages. 1987.

Vol. 1252: T. Kaise, Représentations de Weil et GL_2 Algèbres de division et GL_n. VII, 203 pages. 1987.

Vol. 1253: J. Fischer, An Approach to the Selberg Trace Formula via the Selberg Zeta-Function. III, 184 pages. 1987.

Vol. 1254: S. Gelbart, I. Piatetski-Shapiro, S. Rallis. Explicit Constructions of Automorphic L-Functions. VI, 152 pages. 1987.

Vol. 1255: Differential Geometry and Differential Equations. Proceedings, 1985. Edited by C. Gu, M. Berger and R.L. Bryant. XII, 243 pages. 1987.

Vol. 1256: Pseudo-Differential Operators. Proceedings, 1986. Edited by H.O. Cordes, B. Gramsch and H. Widom. X, 479 pages. 1987.

Vol. 1257: X. Wang, On the C*-Algebras of Foliations in the Plane. V, 165 pages. 1987.

Vol. 1258: J. Weidmann, Spectral Theory of Ordinary Differential Operators. VI, 303 pages. 1987.

MIX
Papier aus verantwortungsvollen Quellen
Paper from responsible sources
FSC® C105338

If you have any concerns about our products,
you can contact us on
ProductSafety@springernature.com

In case Publisher is established outside the EU,
the EU authorized representative is:
Springer Nature Customer Service Center GmbH
Europaplatz 3, 69115 Heidelberg, Germany

Printed by Libri Plureos GmbH
in Hamburg, Germany